The Bird Singers

The Bird Singers

How Two Boys Discovered the Magic of Birdsong

Jean Boucault and Johnny Rasse

Translated by Katia Grubisic

GREYSTONE BOOKS
Vancouver/Berkeley/London

First published in English by Greystone Books in 2025

Originally published in French as *Chanteurs d'oiseaux*, copyright © 2023 by Les Arènes, Paris. Published by special arrangement with Les Arènes, France, in conjunction with their duly appointed agent 2 Seas Literary Agency.
English translation copyright © 2025 by Katia Grubisic

25 26 27 28 29 5 4 3 2 1

Greystone Books Ltd.
greystonebooks.com

Cataloguing data available from Library and Archives Canada

ISBN 978-1-77840-183-1 (cloth)
ISBN 978-1-77840-184-8 (epub)

Editing for English edition by Paula Ayer
Proofreading by Meg Yamamoto
Jacket and text design by Javana Boothe

Jacket illustration by shuoshu (landscape), Gulay Erun (two boys), Piyanat Nethaisong and Gyzele (birds)

Printed and bound in Canada on FSC® certified paper at Friesens. The FSC® label means that materials used for the product have been responsibly sourced.

Greystone Books thanks the Canada Council for the Arts, the British Columbia Arts Council, the Province of British Columbia through the Book Publishing Tax Credit, and the Government of Canada for supporting our publishing activities.

The publication of this work was supported by the cultural service of the French Embassy in Canada.

Greystone Books gratefully acknowledges the xʷməθkʷəy̓əm (Musqueam), Sḵwx̱wú7mesh (Squamish), and səlilwətaɬ (Tsleil-Waututh) peoples on whose land our Vancouver head office is located.

To watch video clips of the authors imitating some of the birds you'll encounter in this book, visit the link in the QR code below, or go to greystonebooks.com/products/the-bird-singers.

To the birds of the day
and the birds of the night,
all the birds who visit us in our dreams.

Contents

PREFACE

A Tale of Two Boyhoods

THIS IS THE STORY of two children from the Baie de Somme region in the north of France who were passionate about birds, two boys who early on developed an unusual talent: they had a knack for mimicking birdsong. They talk to birds, and create kinship between the avian world and our own.

We wrote this book together and separately: each chapter offers one point of view, and our voices answer each other. To help identify who's holding the pen, a drawing of a bird prefaces each section—a herring gull for Jean, and a blackbird for Johnny. Why? Come find out!

Herring Gull

I WAS STANDING on top of a desk in a lecture hall, screeching, screaming like a gull. A herring gull. It was frosh week at the Faculty of Pharmacy at Amiens. Two hundred and fifty frightened students were staring at me, and another fifty or so alumni were up at the front, cap-and-gowned and rowdy. One of the initiation rituals was to find out who among the first years could scream the loudest. Everyone took a turn, and now I was up. I shrieked at the top of my lungs: at first the sound was human, then my voice grew louder, louder and higher pitched, rising, and out came the bird. It lifted into the sky: I was a herring gull, screaming, soaring out over the crowd.

They loved it. The students started cheering, and the judges approved, christening me, with the limited ornithological knowledge of pharmacy students, the school's official seagull. The year was off to a good start. I'd always been a little afraid of others, of humans, or rather, of crowds—an odd fear, perhaps, for someone who was six feet tall and who wore size-fourteen shoes in eighth grade. I was always the kid who stuck out, literally, though I could never stick up for myself. As soon as any situation escalated, I didn't know how to react; there was nothing I dreaded more than being put on the spot like that. But this time, the gull saved me.

What was I even doing here? Was I really going to become a pharmacist like my father, when my childhood dream had been to look after birds?

Down in the street, a black redstart rattled and a gray wagtail answered, slowing the finish, just like the one that used to nest in the playground at my elementary school in Arrest, where I grew up.

Where it all began.

School Days

MY OWN STORY BEGAN in Arrest too, a tiny village of no more than nine hundred souls, three miles from the sea as the crow flies. It's not quite the sea, of course: what we called the sea was the bay, the Baie de Somme. Our village was surrounded by lush groves and floodplains, and a brook we called the Avalasse cut across the town. In the evenings, gulls flew back and forth over the stream, answering the high-pitched melody of song thrushes, an improbable dialogue between land and sea. Village elders say that, long ago, horses would stop in Arrest to eat before setting off on their way to their summer or winter pastures: the place was an oasis of sorts in the pastoral trade of yore, and so the hamlet was named Arrest, which means "stop" in Picard, the Hauts-de-France regional language. Arrest has a square, a post office, a school, a butcher shop, a soccer stadium, an old-fashioned café. On Rue de Catigny, the main street, was the pharmacy, the only one in town.

The pharmacy's big glowing cross cast a green blinking light into my childhood bedroom a block away. I slept fitfully, scared of shadows and childhood ghosts, but the green strobe helped lessen my fear of the dark. Gradually, the canopy of the fruit trees in the yard would begin to whisper of feathers, small bodies, and wings: the birds soothed me to sleep. When I woke, I drew them frantically, my guardian angels rendered on paper. I gave them to my parents as gifts. Each bird cast a spell over my dreams and lulled me away from the dark cloud of anxiety. They watched over me as I slept, nestled by the winged beings my father seemed to love more than his own son.

Every morning, I would do my times tables and memorize my poetry quickly, and every day I set off to school, a fifteen-minute walk away, with my mother and my brother. "Bonjour, madame": I heartily greeted the mail carrier, a neighbor, people coming out of the café or the church with my best smile, making my mother proud.

We got to the schoolyard, with its tall linden trees and the bustle of parents and children—walking or in cars, rushing, carrying their books, shy or proud. There was shouting and there were tears—and suddenly, three loud claps stopped the children in their tracks. Like little ballerinas, they lined up class by class.

I loved that school—my teachers, and all the activities and outings, especially those to the Parc du Marquenterre in the nature reserve, and the Maison de l'oiseau aviary. I knew everything about birds, and those field trips were my time to shine. But every time I tried to show off my knowledge to my classmates and the grown-up chaperones—proudly answering, for instance, that the common shelduck is a striking, multicolored duck that hides and nests in rabbit holes—some voice would pipe up: "It's crazy, Johnny's like a bird! But Boucault is even better..."

How could anyone be better than me? I knew everything about these birds: their feathers, colors, songs, beaks, behavior. But the whispers at the back of the room were going strong: the Boucault boy could apparently name any bird, in French or in Latin, he knew all the new nomenclature, and he even claimed to be in touch with the region's leading ornithologists.

I knew exactly who the Boucault boy was. He was the pharmacist's son. He seemed huge to me, at least three heads taller. He was two years older than me, in fourth grade. He hung around after school with his backpack pulled tight to his shoulders, like a parachute.

Sometimes I would see him walk by the soccer field near my house. He looked so clumsy in his rubber boots that his nickname came naturally: we called him Boots. He didn't play sports, and strode across the field without so much as a glance, showing no desire to chase a ball. For me soccer meant freedom—it was about sharing and community, about teams that came together on the fly, with no age divisions, where little kids could play against sixteen-year-olds. I was lucky, I was good at soccer, and it helped me make friends.

It was seven o'clock at night. My father was home from work, and the smell of sheep in our house was almost unbearable. Even though my mother shoved his work clothes into the laundry as quickly as possible, it was no use: the flock had permanently seeped into the walls of our house. My father was a farmhand for a wealthy landowner, a nobleman who owned cattle, land, and even a château. The smell of the farm stuck to his skin. Some might call it the smell of the working class, of physical labor, of poverty—a mix of sweat, exhaustion, and animals. To me it smelled like evening.

A Letter From the Prefect

OF THE THREE APPLICATIONS my father had submitted to set up a pharmacy, the one for the village of Arrest was accepted by the Somme prefect. I was two years old when we left Toulouse and its chocolatines to land in what was to become the charmed world of my childhood: the area between Saint-Valery-sur-Somme, Le Crotoy, Boismont, and the Rue de Catigny. I quickly found out that the prefect's decision had landed me right under the major maritime bird migration route for the Western Palearctic (which refers to one of the eight biogeographic regions).

We lived right below the flyway. In books, we always looked down onto the map, but I only had to look up to watch the birds fly down to Africa before winter, and back to the Arctic Circle in spring.

The Rue de Catigny was a mile-long road that was like a little village itself; that was where I grew up, happily. I bore the honorary distinction of being the pharmacist's son.

Everything started on the road to school, especially on the way home. My parents agreed that I could bike or walk, as long as I didn't dawdle too much on the way. The trip back from school mercifully ate into my homework time, and it fulfilled my need for observation.

I snooped around in farmyards and inspected everything everywhere. At that time of day, the sky was a peculiar spectacle: a great cloud of gulls flew down in a V formation at sunset into the bay to spend the night. As soon as the birds spotted the shore in the distance, a huge commotion rose up—first the lead bird started in, then the rest of them, a warning to old potters or fledglings bringing up the rear that the night's bed of sand and sea-foam was close by.

Every evening, I attended that ceremony, watching something sublime and otherworldly take place as the birds gathered in the autumn dusk. That was where it first happened, between the Lelong pasture and the Aliamet house.

I was out walking. A few seabirds slipped by over me, squawking. Startled, I turned and watched the thirty or so birds go by. Without knowing why, I started squawking too, a cry of both astonishment and entreaty, a call that burst out of me, a song I somehow knew. The song of the birds that had just flown by.

The flock immediately veered left and headed east, wings tight. They flew over the farm and back over me, emitting the characteristic little quack of a gull seeking contact. Then they were off again. Once more I called, and they came back. They were answering me! My voice and theirs, in unison. It was just me and the birds out here. I was flying, carried along with them, drawn by the thread of sound that bound me to the sky.

Higher up, a much noisier flock distracted my winged friends and poached them from their new, grounded brother.

I could sing like a herring gull! From then on, every day, rain or shine, as soon as school let out, I headed to my meeting with the birds. Over time, I tried out different calls depending on age and hierarchy, even the squeaks of hatchlings begging for food.

The days got longer, and my after-school gull schedules didn't work anymore, but I discovered more songs down in the street—geese at the Monchauds', a rooster and pigeons at Tcho Pierre's, a blackbird or thrush in the Gardins' cherry orchard... There were roosters and hens at the Blondins' and the Seigneurs' houses, but they were unruly little chickens that didn't take any crap from anyone. At 56 Rue de Catigny, there were six hens and three roosters. The Sussex chickens were out in the field, their white plumage standing out against their small black collars and bright red crests. They looked like aristocrats strutting their stuff out in the grass. They were the prettiest hens in the neighborhood, and they knew it.

I let out a few friendly guttural *pehoos.* The roosters glared down their beaks at me. Excellent.

Stay at the fence for a few minutes, make them forget you, watch, gradually become a rooster. Straighten up. Lock your eyes down in the bottom of their sockets, lift your neck to give it that mobility and the extension that goes along with the roosterly gaze. The body tilts. Weight on the balls of the feet, knees slightly bent, rump bulging. The tail that would have sprouted in seconds feathers to life.

I was ready. The Arrest rooster jam was about to begin.

Flap your wings five times, here comes the accelerando,

one, two, three, four, five, slap your hands against your thighs, wait seven seconds and then do it again, like a conductor giving the beat and making sure everyone is together. Then, and only then, can you start the cock-a-doodle-doo.

The response there in the yard was immediate. First a Sussex rooster crowed, then a bantam, jealous of its tenor friend. From one farmyard to the next, Sussexes and Marans, Wyandottes and Silkies, the whole village became a henhouse band.

My fowl orchestra.

A Shepherd and His Flock

AS A CHILD, I spent my Saturday afternoons with my grandparents. Their house was an old farm, with sheds, hens, roosters, a big vegetable garden, and rabbit hutches. There was also a fenced-in pen that was particularly tempting, not least because it was strictly off-limits to us, chained and padlocked shut. My grandmother watched our every move from the window, a vigil against the slightest misbehavior. The rules were clear: we weren't allowed to set even a toe off the flagstones. The rhubarb patch and rows of potatoes were forbidden. Our grandmother's nickname was swiftly and aptly coined: the Foot Police. She scrutinized the soles of our shoes every time we came in. After the foot check, we got a treat—a piece of homemade clafoutis. It didn't matter if it tasted a little tart or too sweet: if it was on the plate, into the belly it went.

As he did every Saturday, my father took a key from above the clock and headed outside, my grandfather following behind. That time, my little brother and I leaped out of our chairs to go with them, only to be stopped by our grandmother, the inspector: "That's for grown-ups!"

"But I'm big," I couldn't help whining. "I'm four feet tall and almost fifty pounds! The nurse told Mom!" I pointed to my brother: "He's the one who's little."

My grandfather grabbed my arm and left my brother standing there. And suddenly there I was, standing before the padlocked gate. With a click of the key, a vast, green Eden crowded with tall trees opened up to me.

My father walked over to a log and grabbed a wide net made from a potato sack. My grandfather stood a little farther away, armed with a stick and burlap. The three of us waited, silent and still in front of a snarl of nettles. And—I saw it, I wasn't imagining things—the mass of nettles moved. *Pshhht, pshhht.* I strained to hear. *Pshhht*... every three seconds. There was definitely something there. My father and grandfather stood by.

"Don't move! They're coming out!" my father shouted.

Move? Not a chance: I was terrified! My heart was pounding in my head and I felt wobbly. I could only hear that one sound, that hissing, incessant *pshhht*, like a metronome. The sea of nettles stirred again. The thing was gigantic! That explained the padlock, then: there was a monster living in the garden, and we had to catch it. My father and my grandfather seemed pretty ill-equipped with their stick and landing net.

Suddenly, a green head appeared, and a golden yellow beak, with a narrow white collar tight around its long neck. I recognized it immediately: a mallard drake—*ch'maillard* in Picard. The duck's defiant stare was a relief. This was the nettle monster? With a yawn, the bird stretched his beak wide and let out a cry—almost nothing, just a discreet hiss, *pshhht*, as if he were satisfied with his own animal beauty. This almost silent little beast wasn't that monstrous after all.

My grandfather began to shake the burlap sack furiously. One little brown head appeared in the snarl of green, then two, five, a hundred heads. This was the monster's body—a hundred mallard hens! For a single male! With their plumage the color of dry leaves and gray-brown beaks, none of them were as flamboyant as the drake. They were the same species, but their coloring and attitude were different: the females seemed more curious, more mischievous.

One duck, with a lighter head than the others, stretched her neck like a stem drawn to the light. She puffed herself up and shrieked with pleasure as she turned to the sun peeking through the clouds. Her voice was clear and powerful: I was hooked from the first note. This was a duck with the voice of an opera singer.

As if they were tuning to concert pitch, a hundred voices rose up in unison. The tall branches trembled. All the ducks started calling at once. It was pandemonium. We couldn't hear each other at all. My father was gesticulating wildly at me to go forward or back, I didn't know, I couldn't hear a thing. My grandfather, meanwhile, was already deaf, so it didn't make much difference to him. Hundreds of beaks lifted up to the sun and sang out together. A cloud slid back over the sun, and the set drew to a close. In the sudden quiet, without a word, my father and grandfather started to plow straight through the nettles. My father waved me along. The drake and all the females thrashed about in a clatter of wings and feathers as they tried to fly away, but it was no use;

they froze before us like frightened lizards. The ducks were trapped, funneled into a corner of the fence. Like a shepherd sorting out his sheep for shearing, my father caught the ducks one by one with the net, grabbing their wings with a swift, reassuring grasp.

I can still see him, bird in hand, the wings clamped together, my father reaching his arms toward the sun, as if he'd won the World Cup. What was he doing? The duck, which had been quite mute, was now singing continuously, bellowing out trebles as much as her breath would allow. A long, loud rattle rang through the yard. The bursts of *queg, geg, geg, geg, geg,* always five in a row, were on par with the greatest songstress. The jury seemed pleased with the surreal performance. My grandfather may have been deaf, but he nodded his approval. "Long call, very pure," my father said, carefully placing his treasure in the sack.

Huddled in the bottom of the bag, the duck calmed down. She stopped moving. I was holding the bag; my only job was to keep it closed so she couldn't get out. I peered inside, but my father brought me up short: "Close the bag, she'll get away!" I wanted so much to pet the duck, give her a kiss, and let her go. "What are you sniveling about? I'm going to send you back to your mother," my father scolded me. "You can't tell this kid anything, he starts wailing."

I was crying at the little duck huddled in the dark at the bottom of the bag. I knew she was scared too. We were both trapped at the bottom of the garden. Fortunately, the next duck caught in the net wouldn't suffer the same fate. Her beautiful marbled coloring promised a unique voice, but she quacked out a lone *queg* so hoarsely that my father let her go right away. I was glad to see her flutter back to the jumble of nettles.

A dozen of the hundred or so ducks ended up in the bag. Beaks open wide, the divas sang with all their might into the sun, a different, distinct song each time. An intense, powerful, gut-wrenching song. Long call, short call, half call, the slow, drawn-out pleading call... This was a veritable open-air audition, a lyrical prize fight, with each contender dreaming of a career as a soprano, a future coloratura. Each hen's role was determined by pitch, length, and rhythm, and the soloists were markedly more powerful than the others, who were relegated to the rank of choristers in the nettle pit.

The casting call among the poplars and nettles lasted almost an hour. Finally, the burlap bag was closed with baling twine. My father left without waiting for me. My grandfather smiled at me, embarrassed. I was amazed by how quiet and unbothered the ducks seemed in the pitch dark of the sack. I imagined them padding all over each other, without a clue where they were headed... Where were they going, anyway?

A Mercedes had just pulled into the yard, and two men in suits stood next to the tractor. My father swore under his breath as he got to the gate: "Shit. Those clots are here already." The exchange was cordial enough—about live decoys, good winds, blinds, the Baie de Somme, duck calls, all in a strange accent. I half caught that my grandfather was boasting to these Parisians about our sublime singing ducks. They had to be positioned just so, taking into account the winds, so they could call to attract their friends. My father's people were famous for raising tamies—the female mallards tied up and used as live decoys to lure wild ducks—and for bagging the best singers in the region. That was my grandfather's job. But after a lifetime of hunting, he'd gone deaf, and my father was his ears now to help him select the best ducks. Without him, there would have been no bag, no ducks, and no fancy buyers from Paris. A hundred francs got them the bag of ducks on loan, with a night in the hunting blind thrown in. The Mercedes started up, taking the men out to the bay. They'd be back tomorrow at noon with the ducks, which would be set free.

When the transaction was complete, my father and grandfather took me to the village of Forest-Montiers, on foot. After a quarter-mile walk along the narrow streets and past the café on the square, we came to an unfenced pasture, acres deep, crossed by a wide stream. Standing in the middle of the meadow, facing into the wind, my grandfather stopped, then tapped the ground several times with his walking stick. He furrowed his eyebrows and, mouth half-open, he whistled, a high-pitched trickle of air passing between his teeth. It sounded almost like a lost duckling. The repeated call was a magnet, and gradually, with each trill, the stream filled up. Hundreds of ducks, previously out of sight, gathered on the water as my grandfather repeated his call over and over.

The birds were all around us now, calm and peaceful. We were in the midst of a tide of mallards, males and females, splashing around without an ounce of fear. My father stood back and my grandfather slowly stepped forward. The ducks moved aside to let him by, the crowd opening up before him. He started making a new sound, slightly different now: he was telling them to follow him. Holding on to his stick, he headed back to the village, the long parade of ducks waddling behind him. He led the impromptu convoy downstream, to clear water and abundant duckweed—a favorite of his flock. Then the bird shepherd backed off, letting the ducks roam.

The village café has a photo of that pied piper moment framed over the fireplace in the front room. The photographer captured my grandfather, his hand on his stick and his cap pulled tight on his head, walking along, with ducks all around him.

My Duck

"WE'LL HAVE AN ANIMAL in this house over my dead body!" That was my mother's predictable response to the question of a pet—a dog, a cat, even chickens… Not that we didn't have any critters at home—our heads, specifically, were well populated. Were my sister and I the lousiest kids in the village? They say the cobbler's children have no shoes. Maybe the pharmacist was using expired lice shampoos.

The village fair was held on the last Sunday in August. The year I turned nine, I was allowed to go alone for the first time. The sound of thwacks, bangs, screeching, and shouts echoed across the grounds. There were shooting games with colorful balloons, a chair swing ride, huge bunches of cotton candy swirled up by an enormous woman with filthy hands. The pièce de résistance, in front of the community center, was a round steel trough, as big as the moon and filled with water.

The trough was surrounded by metal barriers to prevent people from getting too close. A gleeful throng waited behind the fences. I slipped through, and between all the legs and elbows, I caught a glimpse of what all the excitement was about: ducks! Three live ducks were bobbing on the water. They'd been taken from a farmyard in the early morning and seemed amazed to suddenly have access to so much clean water at the height of August.

"Three rings for five francs, three rings for five francs," cried Tcho Pierre, who was running the show. Everyone waited their turn to toss the rings into the makeshift pond, with the approximate skill of a drunken surgeon. If one of the rings miraculously ended up around a duck's neck, the lucky thrower got to go home with the animal, cheered on by the crowd. Lucky or not, it took about a hundred tries to snag a duck. When a duck was won, a replacement would get brought out from a nearby wicker box, to keep things fair.

I watched the ducks. One of the three, a magnificent mallard drake, with a head that was beginning to take on that iridescent green, had caught on to the gimmick. He was floating so high in the water that he looked like he was standing up. He was beautiful; he was the one everyone wanted, but he was determined not to be hauled away by a stranger. At every throw, he would nod his head into the water, and the ring splished dispiritedly against the surface.

Suddenly, I heard our family name. My heart stopped. Pierre was calling my parents up. They'd come to the fair after all.

"Come on, madame! Three rings for five francs! It's for a good cause: the money will help cover the cost of travel for the soccer club."

My mother stepped forward gamely and paid five francs for her three rings. I hid, expecting everyone to laugh. Yet she confidently threw the orange ring, and then the blue. They didn't even hit the trough.

"Throw harder!" the crowd egged her on.

The last ring, a red one, bounced off the edge and spun along the rounded curve of the corrugated steel like a tightrope, careening into the water. The ring sank, then rose straight up again and flopped idly at the surface of the water. The splendid mallard, which hadn't anticipated the ring's meandering course, dove as it usually did. But then, before the astonished crowd, the bird lifted its head, accidentally hooking the ring.

She did it! We'd won! They handed her the duck, which struggled against her like some wild reptile cousin with claws, a beak, and powerful wings, intent on punishing the lucky winner. I was overjoyed. I rushed to my mother, grabbing the duck and grasping its wings as I'd seen so often on farms. I held on tight, announcing triumphantly to anyone who would listen that we would never eat it. In the end, the schoolteacher's husband offered to take in the duck while we got organized to house our new pet.

In the evenings, I pored over ornithology books. The mallard, *Anas platyrhynchos*, order Anseriformes, family Anatidae. They weigh between a pound and a half for females and three and a half pounds for males. Mallards are the most common dabbler. Males and females are considerably different: the drake has bright green plumage on his head and neck, while hens are brown. Their life expectancy, I learned, can be up to twenty-nine years, a fact I decided to keep to myself.

The following week, we got ready to bring home the duck. My father's job was to buy stakes and fencing for a pen, and sand, gravel, and cement to make a little pond. After school, I ran over to pick up my duck. As I came near the barnyard, I only had eyes for him among the thirty or so ducks frolicking in the mud puddle that served as their bathtub. Already

I knew him like the back of my hand—the look in his eyes, with his head tilted watchfully, that little double tail curling at the tip, the striking blue-purple reflection on his wings.

"That's the one!" Someone pointed to a loser of a mallard, which didn't even seem to know that he was part of the big Anatidae family.

"No, no!" I refused adamantly, not without causing some embarrassment to my family. "No! No! No! I don't want just any duck, I want my duck."

My parents talked it over, and I got my duck. And, with a few extra pennies, I even got to take home a pretty female, because as everyone knows, life without a girlfriend is boring.

I spent hours watching my ducks from the kitchen window, spying, learning their little secrets. The female was always out front, in charge of the couple's daily destiny: she decided when it was time to swim, and when and where to rest. The male didn't argue, and got into the habit of following her. Once or twice a day, however, slipping into the pond after his mate, the male seemed to puff up. He would stand upright on the water and, with his beak pressed against his chocolate-brown chest, he gave a barely audible, high-pitched whistle. So much energy expended to produce such a pure but mediocre note.

Yet the sound was like a love potion: the female circled the male and flattened herself like an alligator at the edge of the water. Then she got up and stood right behind the male, tilting her head in his direction and emitting a frantic, nasal *queg, geg, geg.* The drake proudly tottered out of the pond, repeating his raspy chirps. She loved him, and he knew it. Reassured for the time being, he let her take over again.

Feather Song

MY MOTHER GATHERED SOME supplies for us for the night: a loaf of bread, the gas stove, a tin of lentils. I was six years old and going hunting with my father for the first time. I had no choice: I was the oldest, and the tradition was that I had to experience the Baie de Somme by night. The expedition, even in the midwinter cold, would have been a dream for my cousins, neighbors, and school friends.

On the edge of a two-and-a-half-acre pond stood a half-buried cabin—the Rasse family's pride and joy, a rare property for a modest family. That was where we were headed. I was bundled up from head to toe, with a balaclava and two scarves.

Low tide. We entered a strange, dark labyrinth. The lamp my father held lit up the ground. Everything was slimy, and the clay made it hard to walk. I struggled to keep pace with my father, slowed down too by the decoys I was carrying, plastic silhouettes he would place in the pond to attract waterfowl. Their strings tangled in my feet and I could hardly walk. I wasn't allowed to complain, but I sneakily lightened my load, dropping a few here and there on the path. Five decoys seemed like enough… Twice, I lost my balance and fell in the mud. My gloves were soaked and dirty. The smell of silt washed over me. I was crying. After a half-hour walk, I finally spotted the hut, buried under a mound of grass. We were about fifty feet away, with a churning stream between us and the cabin. I was shaking with fear at the final obstacle, a slippery makeshift bridge made from a railway tie. I was afraid of heights, and didn't dare look down; I couldn't move. My father was already far ahead, annoyed: "I'm not going to carry you, am I?" I could hear the power of the current under my little boots. I was petrified. I could only hope the ropes on the decoys wouldn't get tangled.

Finally, there we were, standing on the cabin roof. Through a rusty iron trapdoor, we lowered ourselves inside a sort of wooden box no bigger than forty square feet. Inside, there were two single beds, about a chair's width apart. It smelled as musty as our cellar at home. The spiders were giant and the bedsheets were covered in old wine and coffee stains. We must have forgotten some decoys in our rush to leave, my father noted. I simply nodded.

Once dinner was over, my father blew out the candle and opened the little casement through which we would watch the birds at night. I marveled at the moonlight and the stars; I'd never seen them so sharply. The cold night was calm and soothing, with the furtive cries of sandpipers and gulls and the delicate lapping of the water.

As I found out, there were rules in the hut: speaking loudly was forbidden, and in fact we could only speak when it was absolutely essential. My throat was sore and I couldn't whisper. Luckily, just before we left, my mother had slipped some honey candies in my pocket. My eyes were riveted to the light glimmering on the surface of the pond. My head was full of questions. "Papa, why do ducks only move around at night? Why can't we hunt shelducks? Why do you shoot birds?" Exasperated, my father just told me to go to sleep. I felt like I was being punished. I snuggled into the sleeping bag. It smelled like home, and I breathed it in as deeply as I could.

Everything there was horrifying to me. We washed the dishes in pond water. The mattress springs stuck out and scratched my back. My aching feet. The bitter cold. The fake birds we strung up to lure others. And the gun. The smell of kerosene from the stove kept me awake. I coughed. My father got angry: "You're such a pain! This is the first and last time you're going to come hunting."

Finally, I fell asleep, sinking into my childhood dreams.

Suddenly, an awful bang tore through my sleep. My whole body was shaking. I was crying, screaming in pain. My father dropped his rifle and rushed toward me, touched my head, and looked at his hands: there was blood on his fingers. I could see the panic in his eyes. I turned my head away from a deafening whistle, and then there was nothing.

When I opened my eyes again, I could see my grandmother. I could see a cardboard box with coffee éclairs in it. I was lying in the middle of a big white bed. A doctor was explaining to us that my eardrums had ruptured: perforated membrane, and Eustachian tube damage. I would probably be partially or completely deaf as a result, he told my parents.

No one before me had ever had burst eardrums in a hunting cabin from a shotgun blast. From that day on, I was the laughingstock of my uncles, and doted on by my aunts—the eldest son, but who must be protected because he's fragile and sensitive. As my mother said, my sensitive ears would keep me away from the hut for good. I couldn't go hunting; my body had decided for me.

We went back and forth to the hospital. My hearing was recovering slowly. Months passed, and spring came, and with it the great African migration. My father loved the bay, and we went back to the hut together, without a gun this time, and out of season. I found myself behind the pipers, with a front-row seat to thousands of birds. Each male had his own mating strategy, using strength, gentleness, and charm, before taking to the skies in pursuit of his beloved. It was a merry-go-round of seduction, a puppy-love ballet.

Settled in the half-buried hut, I was mesmerized by an unfamiliar sound shearing the air. I stopped, immobile, and held my breath for fear of losing the miracle of that sound washing over me. I could make out a whistle that had nothing to do with the tinnitus I'd been living with. It sounded like... air moving through space. I could hear birds' wings! I was discovering the tiny song of feathers. Closing my eyes, I could imagine each species before I saw it, just from the frequency of the beating wings. An invisible realm was opening up to me, at the edge of what could be perceived: the infinitely audible world.

Duck Games

AS SHE DID EVERY WEEK, the mail carrier brought *Le Galibot*, the free classified newspaper. "Village of Lanchères estate sale: 10 ducks for sale, 500 francs the lot." I called the number in the ad, but the lady who answered thought it was a hoax. My father called back; come around and see, she said.

It was about a ten-minute drive to the southern shore of the Baie de Somme. We pulled up to a run-down wattle-and-daub house. There was a pond in the small farmyard, with a big cage and ten colorful ducks. My father made the deal, and now all of a sudden there I was with twelve ducks. My duck didn't know which way was up. There were three drakes among the new ones, and I didn't think he was too happy about that. Then there was this little all-white duck, winsome as all get-out, with blue eyes and a yellowish-orange beak starred with little black marks. She was my favorite.

When I was in fifth grade, I left school in Arrest for a new school in the working-class town of Friville-Escarbotin. My new classroom in the city smelled different from the countryside. There had been no school cafeteria in Arrest; all the students went home for lunch. When they came back in the afternoon, the hallways filled with the delicate aroma of fried food and mustiness. I took it all in. I could tell who'd been lucky enough to feast on the best thing humans have ever done with potatoes: French fries!

Fries: the very word was forbidden at home. They smelled bad, and they were bad for your health. I savored the scent of the forbidden treat, which whetted my imagination as well: I thought I could smell on Benoît or Pascal the difference between traditional French fries and fries cooked in shortening.

Eventually, once a week, my sister and I were allowed to have lunch with Monique, the farmer's wife. What a revelation that was. Not only did we get the smell, we also got the short-order music. Bubbling at first, the oil would rise to a sizzling crescendo, reaching its apotheosis with the rhythmic shaking of the basket in the fryer. And, at last, a taste of the divine. Cast-iron fryer, double-fried, hand-cut… Monique could work wonders with Bintje, that queen among tubers. On the days we ate at their place, we smelled like fries, just like everyone else. It was comforting to belong.

At Friville-Escarbotin, everyone ate at the cafeteria—mushy chicken, noodles, duplicitous cordon bleu. The classroom, meanwhile, reeked hopelessly of dust and white glue. Only Karim didn't eat at the cafeteria sometimes. On those afternoons, he smelled amazing.

"Do you have any ducks?" the school principal asked me one day.

"Yes, I have eleven." By then one had flown away.

"Would you lend them to us for the duck games?"

"Duck games?"

I found out there was always a race at the end of the school fair. The concept was simple: all the students from kindergarten to sixth grade would sit in the schoolyard in two rows facing each other, with about fifteen feet between them, creating a straight sixty-foot track. Each of the teachers would have their own duck. Bets were placed. Every contestant had to follow their duck to the finish line. The first three ducks across the finish line won, and whoever predicted the three winners went home with prizes donated by local merchants.

That year, my ducks would be competing. All I could think about was how to make sure my teacher won. And the second-grade substitute teacher, too; she was really nice. Every night, I worked to train the white duck, and another duck, a blond-feathered hen—walking in a straight line, maintaining their pace and speed without getting carried away. I used worms and lettuce to reward them when they did well, and my duo got better by the day.

Finally, it was the day of the school fair. By two o'clock, my whole family was there. My ducks were in their crate, ready for the closing event. The late June sun was beating down, and my ducks were hot, too hot. They needed shade. I found a cool spot outside the school, between a cedar hedge and a brick wall, and set their crate there. My friends were playing and laughing, but I was stuck with my ducks. I didn't want anyone to come near me. My birds were doing better, but I couldn't leave them. I sat on the crate and watched them between the wooden slats. They blinked and fell asleep. I did too, and even missed going up onstage to sing with the other fifth graders. The music stopped. I could hear the crowd chanting, "Ducks! Ducks!"

When I got to the playground, carrying my crate, all the students were sitting on either side of the racetrack, and they watched as I made my way along the honor guard. I opened the crate and handed the white duck to my teacher and the blond one to the substitute teacher. The other teachers got the remaining ducks. Everyone put their duck down, and the countdown began… On your marks, get set, go!

All the kids were screaming. The white duck, my teacher's, was the first across the finish line, by seconds, followed by the substitute teacher's blond duck. A drake wrangled by Monsieur Hannock was third. Exactly what I'd planned! No one noticed that the race results had been rigged by careful training and patient practice. I'd done it, my teacher had won!

Unfortunately for me, however, I had forgotten to place my bet.

Shadow Puppets

MY FASCINATION WITH the colors and shapes of birds led me, at the tender age of seven, to delve into a book that had previously been forbidden to us children, a book tucked safely in a drawer in my parents' bedroom, away from the rest of the family. As the oldest, I was the only one who was allowed to flip through it. I was careful never to move it, and I spent hours with it.

Snuggled in the soft cotton sheets on my parents' bed, I looked at all the feathers, colors, beaks, and silhouettes of each species to get to know them better. The illustrations in the thick guide brought me to the Far North, the Trevor Boyer drawings of Palearctic Anatidae really coming to life. The spectacled eider, Barrow's goldeneye, and hooded merganser were storybook characters to me, my imaginary friends in that family haven.

One night, I stayed up later than my brother and sister. Our house was cloaked in winter twilight, and there was something melancholy in the air. We often had power failures, and the candlelight in the kitchen that evening caught the mood. My mother was asleep but my father lagged behind, as if he were haunted by secret memories. Suddenly, he stood up, firm and determined. He brushed past without seeing me and opened a cupboard behind me.

Groping in the half-light, he took out a plate, its contents covered with a white cloth like a shroud, and placed it in front of me. Every gesture was slow and subtle. The plate was enthroned between the candles and the remnants of supper, while shadows danced on the walls.

The vigil felt baroque. The only sound in the house was the crackling wood fire. The silence induced contemplation. The chiaroscuro of the room evoked Caravaggio, or a Rembrandt still life, with the shrouded plate at the center of the painting. My father reached out his hand and lifted off the first layer of cloth, letting it fall away. When he removed the second, I saw two ducks I recognized from the book in the bedroom: two wigeons, *Anas penelope*, later reclassified as *Mareca penelope*, a male and a female, also known in French as whistlers.

I was fascinated, I couldn't stop staring. I had never seen any wigeons so close before. Every detail of their anatomy was sublime: the webbed feet for swimming, the color of the beak and its shape, the flat, white belly designed for gliding over the water. Until then, this dabbler had been just a picture in a book we weren't allowed to touch. I could almost feel the painter's obsession with his model. Even though the muse here was limp, inert—a bird washed up on a beach—its beauty soaked into my young eyes, imprinting itself on my retina.

There are evenings when poetry dances with death, when what is sensed and what is imagined improvise a little piece of eternity. The kitchen was a candlelit puppet theater. Curled in my chair, I had a front-row seat, ready to enter the world of make-believe, of fancy nights out at the theater. I waited. My father had stepped back into the half-light, across from me. The kitchen table between us was the stage, the set was ready, and the curtains folded back on the two lifeless creatures I was waking by candlelight.

The show started. Two arms reached over the table's proscenium. My father was the puppeteer, and the birds were an extension of him. His hands caressed their heads, his fingers running down their fragile, slender necks. While the show lasted, the two wigeons were resurrected, as if the stage had the power to bring them back to life.

The puppeteer stared at me, as if telling me to look at the birds. His eyes, which I struggled to see in the half-light, became two lonely, distant lights at the back of the stage. The illusion was complete, with only the smell of the burning candles drawing me back into the room for a moment. The spectral light of our Elizabethan theater flooded the stage. I was completely taken in by the charm; I believed utterly in these two avian ghosts.

The male wigeon, its neck stretched taut, revealed a rusty gradation from the top of its head, from lighter to dark, stopping sharply at the collar's contrasting ash gray. The reddish head and neck and the yellow forecrown are distinctive features of male wigeons. On his back, a tangle of plumage looked like a waterproof bespoke suit. The puppeteer's gestures were precise and delicate. He mastered the slightest articulation of his winged puppets, miming a pose, their jittering, searching for food, grooming.

My wooden chair creaked.

"Don't make a sound," my father whispered, as if we might disturb the ducks' embrace. "Look, look at them." His voice was a distant reminder of his presence, which I had all but forgotten. The puppeteer began to narrate the scene, though we were clearly in the realm of art rather than scientific nomenclature or anatomical explanation. He was the birds, he embodied them. He improvised a love scene, the reddish-brown female clinging to the male's outstretched body. *Whee-oooo*: he launched into his love song. *Whee-oooo*, my father repeated, a soft whistle, followed by the duck's characteristic rattle. This mating dance, a few days shy of my seventh birthday, was the first one I would remember.

The deed done, a new act began; scene change. My father's hands flew with a different kind of energy. A miracle was happening here, as if two birds weighing about a pound each had flown into the candlelit kitchen.

Migration stopovers are often short, and departures hasty. The ducks are exhausted from the hundreds and thousands of miles they've flown, and they're so fragile that they're easy pickings for predators. There is danger everywhere: seeing migratory birds calm and at rest is rare. They never stay very long. But the journey is so long and so hard that they have to take breaks. In that moment of silence and stillness, with only the barely perceptible popping of the flame slowly consuming the wax, my chair squeaked, and the puppets, peaceful, started awake.

You could see the tension course through the couple. The birds heard the danger, felt it, and called out a warning. The female, more alert than the male, immediately decided to flee, and after a few seconds of turmoil, she took off from the kitchen table like Columbine leaving Harlequin.

According to the laws of nature, the female sets the pace—a behavior that has helped preserve the species. If a pair were attacked by a peregrine falcon, for instance, it would be up to the male to sacrifice himself; he would give his life to ensure the survival of his kind, allowing his mate to fly off with their young.

That night was a lesson in Anatidae flight and skill. The remiges and rectrices, scapular feathers, the glimmering green and white of the wings... The bird glided by again and again, stopping, drifting gently from left to right as if guided by gusts conjured up by the director, showing off with a few somersaults in midair.

The show came to an abrupt end when the power came back on, the bright lights signaling bedtime. My father carefully put away the ducks. But the memory of that ballet lingered, so much that I too felt like I was being held aloft on a feather.

I was still fascinated by the forbidden book, and decided to draw every single bird in it over the next few years. My masterpiece was the harlequin duck, which my parents framed and hung in their bedroom.

Registration

ONE WEDNESDAY AROUND EASTER, I was spending the morning at a neighbor's farm and they invited me to stay for lunch. They were making chicken. I knew which one it had been; I knew all the animals on the farm. This was a young rooster, who had started pestering the hens, according to Jean-Pierre, Monique's husband. *That's too bad*, I thought; he was much prettier than the old rooster, a bird I didn't trust at all. That one looked at me like he would pounce on me someday, spurs out. Though I must have been a pain too, come to think of it: I used to hide and amuse myself by pretending to be another rooster coming into the pen, laughing as I watched him look around and frantically try to round up his hens.

My hosts knew that I loved the carcass and the head. I played with my food, taking apart the bones one by one, disassembling and reassembling my chicken. They really are architectural marvels, the skeleton and muscles intertwined with tendons like pulleys. I eventually learned the names of each part, but right then, at the table, I was at one with the thing before me. How incredible it is to feel like you're seeing through the feathers, really seeing the body, grasping the animal's movement, deciphering its posture: the neck laxity, bipedal motion, wings folded or extended. I think the neighbors found it funny. I started singing, bringing the chicken back to life on my plate.

After the meal, they turned the channel to France 3 Picardie. We didn't have a television set at home. Screens were bad, or so my French teacher always said. So I sat there like a flycatcher, swallowing the images. Suddenly, two men came up on the screen. They were whistling and talking. They looked old to me. It was a report on the Festival de l'oiseau d'Abbeville, a bird festival in a nearby city. According to the reporter, the smaller of the two men, whose name was Hélios, was the winner of a bird-call competition, and the taller man, Zorro, was runner-up. One had done a gull, the other a cuckoo. That was it. The news moved on to the next story. No one at the table reacted at all.

We had a last drop of coffee, watered down for me. The television was turned off, and we left the table. I decided not to spend the afternoon plowing with Jean-Pierre after all. He was surprised and, I think, a little disappointed. He liked it when I went with him; it was a nice change from the monotony of being out on the tractor alone for the whole day. I usually loved spending my afternoons in the cab watching gulls dive-bomb in to grab worms unearthed by the plow. Despite the noise of the machine, the hundreds of gulls circling overhead made me so happy. I loved watching from afar as the whole flock got drawn in: when a gull sees another gull dive down toward a piece of farming equipment, it follows, and all of a sudden there are gulls everywhere, coming in from unbelievable heights. It's magnetic, dizzying. They have to decelerate so fast and cut extraordinarily sharp turns to get back into the wind. Sometimes the sound of the wind in their feathers is louder than the engine. As the number of birds increased, the shape and flapping of the wings would become more visible from a distance. If our tractor was the only one in the field, the swarm we drew was something else.

Jean-Pierre once told me that during the war his father would put worms on hooks behind the plow to catch the gulls. It was the only meat they had in winter; that was how they survived. I didn't know what to think. Was the story even true?

In early April, the gulls are beautiful with their chocolate hoods, and many of the previous year's young don't yet have their mating colors. But that afternoon I was going to miss the show. I had better things to do. I hurried home. My parents were taking my sister to the conservatory in Abbeville, and I had to go with them. They were surprised to see me back so soon. I made up some excuse to tag along.

Abbeville was incredible: there were birds everywhere in town—bird posters, store windows overflowing with bird designs, flower beds shaped like birds... Even the gun shop seemed to have joined in; there wasn't a single rifle left in the window, just carved wooden birds. At the hardware store, stuffed birds sat on the shelves between the pots and pans. I stopped in front of the big toy store, where there were bird stuffed animals, bird Lego, bird costumes. A tall sign announced, in capital letters, *Winner of the Bird Festival competition.*

I went inside and asked a salesman about it. He was the manager, he told me. They'd won the prize for best window display. He had also attended the previous Saturday's bird-call competition. It was a riot, he told me. Apparently, it was held at the Ponthieu municipal movie theater. I ran off; I knew where that was. They had monthly screenings of *Connaissance du monde* international documentaries. Sometimes the films were about a country that caught my imagination, and I would badger my parents to let us go to the screening and the ensuing discussion. My sister and I were often the only children. The speaker would recount his trip, sometimes talking about birds. We saw Eric Bompard, Haroun Tazieff, and other solo adventurers (though you've got to wonder who was shooting, if they were really traveling alone). The images were so real and the stories so vivid and memorable that I couldn't really believe I'd never actually been to Spitsbergen or to the summit of Mount Etna.

When I got to the movie theater, there was still a poster up for the bird festival, showing a flock of northern pintails. The lady at the box office didn't know anything and suggested I try the tourist office. It was right across from the town hall. I went inside. A woman started to give me a telephone number and finally changed her mind and just dialed it herself, handing me the receiver.

"Hello, my name is Jean Boucault, I'm ten years old and I'd like to enter the birdsong contest."

"With or without a whistle?"

"Uh, without."

"I've put you down. You're the first to sign up for next year."

I had to wait a year?!

I could do herring gulls, and a bit of the great tit's *tea-cher, tea-cher.* At any rate, they answered back, especially the males. And the common cuckoo—just the typical double or triple note, as well as the female's monkey-like racket; I thought I did a better job than what I'd seen on television. And the rooster, obviously. I had to work on my passerines: I had all the melodies in my head, but my whistling technique for the high-pitched tones came out as a muddled hiss, like I couldn't quite get rid of that puff of air. Maybe I shouldn't have sucked my thumb for so long, or maybe my broken tooth was the problem.

A few days later, I came across an article in the *Courrier picard* about that year's competition. Hélios was a professional cabaret whistler in Paris. One day, he said, he was so frightened during a thunderstorm that a super-powerful whistling sound reflexively came out. He won the competition with his cuckoo and nightingale. Zorro, on the other hand, was a fisherman in the Baie de Somme, and an expert in bay-area birds. He could even imitate the little Baie de Somme train! He must have been amazing. The train station was about six miles away from our place, and we could hear the whistle when the wind shifted northeast. In the fall, the sound of the train was part of the soundscape for thousands of migrating thrushes and finches flying over the village.

"The Ponthieu movie house still echoes with the cries of oystercatchers, curlews, and wigeons," the article concluded.

Curlews! That would be my first big challenge.

From that day on, my sister's piano wasn't the only sound ringing through the house. I had a competition to get ready for. I gave up my flight training: for a long time, I'd thought that if I just kept at it and flapped my arms every day like a duck finding its wings, at some point my body would lift off the ground. But I had to face it, the only time I felt like I was flying was when I was doing bird calls.

The start of the contact call was fine: I stretched my mouth forward and made a clean *oooo* sound, then subtly vibrated my lips. But no matter how much I tried, my tongue would slip every time and instead of that rising *ooo* note, I managed barely a *y-oo* or at best an *ee-oo*, never the exact sound. How could I not get that last damn note? How could I not become one with the sound? I tried everything. I'd heard them all winter long, a little flock of about fifteen birds in the pastures along the Avalasse River. Early in the morning, from across the way, it sounded like a flock of gulls, but at the very last second their prominent beaks would peek out, or the powerful cries would reveal the taunting curlews: *Surprise! Gotcha, we're not gulls after all!*

But the first week of March they headed back north, leaving me there like a fool, signed up for the competition with the herring gull, great tit, and curlew. Why hadn't I chosen the whimbrel instead? It would have been so much easier. All you had to do was start with the curlew and then whistle with a bit of a laugh, kind of like whinnying, beginning in a trill and getting faster and faster before stopping short on a clean, open note. The last time I'd heard a whimbrel was the August before, though; they were probably all the way in Mauritania by now, and they wouldn't be back in the Baie de Somme for another month at least, on their way to Lapland.

Help!

"Go see Rasse," Jean-Pierre told me. "He spends all day out on the bay with his sheep. He sees lots of curlews. Maybe he can help."

And so, on his advice, I showed up on Johnny's doorstep. In the kitchen I found his mother, father, and the television set. Everyone was staring at me.

A Visit From the Boucault Boy

AT A QUARTER AFTER SEVEN one night, there was a knock on the door, shy and awkward. I was upstairs with my brother and sister, and we stopped whatever childish game we were playing to make our way quietly to the landing. Worry was written all over our faces: my parents weren't expecting anyone, and the knocking on the door broke the stillness of the family evening.

The knocking grew louder and more insistent, and my father finally decided to call out a stern, questioning, "Come in!" The latch dropped and the door opened halfway. A green boot appeared, then another. I heard my mother get up, leave the kitchen, and go into the hallway, offering up an extremely kind hello, as if to apologize.

On the carpet in the hallway, where my father's dirty boots still stood, I could make out a spindly but not unfamiliar figure. My father, in an audible whisper, asked my mother, "Who is it?" Before my mother could answer, a child's voice interrupted:

"I'm Jean, the pharmacist's son."

"Don't just stand there, come in," my father told him. And the pharmacist's son strode into the kitchen as if he already knew the place.

My brother and sister went back to their game, while I slid down a couple of steps to listen. Jean's reputation preceded him; he hadn't come by that night just to say hello. Even my mother, outwardly calm, sensed that this was an important moment. I could feel her concern reach me through the wooden partition between us. The scent of soap wafted up, and the smell of the house changed—probably the detergent and fabric softener on Jean's clothes—drawing me down a few steps more.

My father began his interrogation. He was cold and authoritarian, as if marking his territory. With great assurance, speaking clearly, Jean recited his answer.

"I'm the pharmacist's son, I've come so you can teach me to whistle like a bird. I already know how to do the herring gull and the turtledove, but I don't know how to whistle, and I heard that you know the Baie de Somme."

His entire being seemed to vibrate when he said that, and I guessed that the Baie de Somme for him was a muse, a mirage, a distant expanse, fascinating but foreign to his young life. Jean didn't belong to the Baie de Somme. His immaculate white shirt collar, small striped vest, and blue pants gave him away: with his impeccable grooming, he looked like he was on a Sunday outing. But the green rubber boots on his feet stood out—a way in, a bridge between his life and his dream of nature.

My father seemed oddly preoccupied, staring at Jean's feet. He could read the marks of long walks across pastures on his boots, treks in the marshes and by ponds, but no trace of the sticky black clay of the seashore. There was no doubt about it: Jean didn't know the bay, he was looking for it and thought he would find it at our place.

My father could be harsh, but that didn't stop the boy. Jean laughed. In three sentences, he managed to show off his knowledge. "I'd like to imitate the curlew," he said.

"*Euch'corlus*," my father replied in the local dialect.

"No, the curlew," Jean said

"Yes: *ch'corlus*!"

"No, the curlew," Jean repeated, "the largest of the shorebirds, with a long, curved beak, from the Scolopacidae family."

A long silence fell over the word *Scolopacidae*. The discussion was over. My father stood up, took a glass from the cupboard, and poured himself some wine, setting the bottle down curtly. Jean, serene as ever, didn't flinch. I could tell Jean was pushing my father's buttons.

"Well, then, pharmacist, let's hear your mewing."

Jean didn't move. My mother jerked her head around, looked my father straight in the eye, and gave him a simple, drawn-out "Jean-B," the diminutive of my father's first name, Jean-Bernard. The game had gone on long enough, she was saying; my father was going too far.

"My husband is asking if you would show him how you imitate a gull," she said, coming to the boy's rescue.

Just like that, the roles were reset, and it was once again a child in front of an adult.

My father nodded. "Yes, can you do the gull for me, please?"

Without a word, Jean's silhouette took center stage in the kitchen. A silence fell over the room.

Sitting on the stairs, I was struck as if by a bolt of lightning, making me tremble from head to toe. The scream was a shard of glass through my body, an inhuman scream, wholly animal. It was so loud that my sister and little brother ran to huddle in my mother's arms, thinking a bird had flown in through the window.

After I got over my initial shock, I scrambled down the last few steps. My father sat frozen in his chair; my mother was dumbfounded. Jean, his arms outstretched like an albatross, had become a bird from head to toe, spinning slowly as if carried by a gust of wind, frolicking in an imaginary storm. He was totally changed. He flapped his arms, as if his limbs were helping him draw a new breath, and again let loose that same divine note, the sound keen and consummate, the same force sending a shiver down our spines. At the beauty of that bird-child, tears began to roll down my cheeks uncontrollably.

Jean didn't stop; he was a gull. In a trance, he sailed between the mountains and the wind. Sounds were pouring out of him, a groundswell. Jean arched his back, launching his song like arrows at the kitchen ceiling, hitting each of us straight in the heart every time.

He stopped, and in the ensuing calm he stared at us with big blue eyes. His face bore no trace of what he had just done, and he was breathing easily. The bird inside him was part of him, it could wake up at any moment, so naturally. It was disconcerting. He was as much a bird as he was human. My father was still stuck in his chair.

"Holy shit, talk about mewing," he whispered.

My brother turned to our mother. "How does he do it, Maman?" She had no answer. She knew we were witnessing a miracle, and that no words could do the moment justice. I remained unmoved, but my gaze was planted in the eyes of the unknown bird, as if to guess its secret.

In the midst of that lull, my father said, in standard French this time, "Come back tomorrow night. I'll teach you how to do the curlew." At the mere mention of the species, the rite of passage was complete, and a pact had been sealed between the pharmacist's son and the bird shepherd, between the boy and the bay.

Bird Lessons

IT WAS A DATE. At seven o'clock on the appointed day, I was at the front door, awaiting the arrival of our *Miaule*, the mewler, as my father had already nicknamed Jean. At a quarter after seven, he knocked. I went to hide on the stairs. My father invited him in. Jean walked into the room with his boots still on and sat down across from my father.

My father started things off. "Can you whistle?"

Jean was quick to reply. "Yes, but not good enough." He explained that he had seen a report on TV about a festival in Abbeville that had a bird-call competition. The idea, as he described it, was that a jury made up of ornithologists from all over the world would listen to and rate bird imitators. The contestants performed on a movie theater stage, in front of a big audience. Each contestant had to imitate three birds. I finally understood why he was here: the day before, during the great gull pageant, Jean had told us he could do a gull call and a dove, but he didn't mention a third bird. So he was here to learn another species, to be able to compete at the Abbeville festival.

Still sitting on the stairs, I waited impatiently for this first lesson. I'd already heard my father whistling blackbirds and waders in the Baie de Somme. His artistry was pretty limited, and so was the birds' response: his technique was to whistle by tightening his lips, a basic whistle anyone could do.

My father asked Jean to whistle. Jean took a deep breath and went for it. Nothing. I was stunned. Not a sound, not a hiss, not a vibration emerged from the unfortunate boy's lips. Even my father, with his archaic technique, was the greatest of singers next to poor Jean, who was desperate, unable to produce a single sound. A gull obviously isn't a singer. It's a screecher, maybe, or a mewler, as Jean's new nickname attested. Jean was a divine gull but a lousy whistler—a far cry, so to speak, from passerines and other waders. He was tone-deaf; whistle-deaf.

Seeing this, my father perked up, launching into an outstanding rendition of the curlew, a long melodic phrase made up of jerks and rolled notes. The curlew is one of the most beautiful of all the sea singers, and a fairly easy bird to call: they respond readily and invite dialogue. When it hears a whistle, the bird approaches in search of its mate and lands a few feet away. Curlews aren't shy, they're sociable and curious. They're not especially concerned about musical quality, either, and were tactful enough to fool my father into thinking he was the world's greatest curlew imitator. Jean seemed impressed, and my father kept singing. Wide-eyed, Jean tried again, to no avail. Nothing seemed to work.

"That's exactly what I want to do," Jean told my father. "How do you do it?"

"If you can't whistle, don't bother," my father replied, suggesting that Jean practice on his own to be able to at least produce a sound before coming back.

I waited a good two weeks for the mewler to return, and return he did, still with his boots on, and as motivated as ever. But that Wednesday evening, my father greeted him with a cold smile: it was Champions League night, and we all wanted to watch the Olympique de Marseille soccer game.

"Have you learned how to whistle yet?" he said, without so much as glancing at Jean.

"Well..." Jean mumbled, hesitating. He could tell he was intruding. "I'll leave you to it, I don't think this is the right time." But instead of leaving, he stayed rooted to the spot next to my father, as if it would be up to him to kick him out. A minute went by. My father spoke again: "Fine, go ahead! Show me."

Like a dunce in front of the teacher, clearly feeling the pressure, Jean threw himself into it. His mouth was clenched tight and he stared off into the distance, concentrating intensely. Every breath seemed to be his last. His first attempt went down in the annals of mediocrity; my father was slightly hard of hearing, and the whistle didn't even register. But from the top of the stairs, I could hear a tiny note graze my eardrum. On his second try, the note reappeared, clearer and more present, admittedly contaminated by a parasitic murmur. But Jean had just whistled, I was sure of it. My father got up, his chair creaking, and asked Jean to hold the note. For an hour or so, with a patience I'd never seen, my father worked with Jean to stabilize the note. That little whistle, like a ray of light through a crack in the wall, expanded and opened up into a second, even higher-pitched note. Each time that note appeared, Jean heard it vanish immediately, as if the joy of hearing the sound lay in its rarity. Yet I could hear two distinct notes, and they were already far more beautiful than anything my father had ever managed.

The Champions League theme song boomed through the living room and brought the lesson to an end. My father, taken aback by how long they had been at it, bolted out of the kitchen. He opened the front door and motioned to Jean to leave as quickly as possible. There was no question of missing the start of a match with Jean-Pierre Papin and Chris Waddle.

The next day, at the same time, there was another knock on the door. Tired, my father got up and shuffled toward the entrance.

"*Chés pas possible, chés encore qui cho,*" he mumbled in Picard: who was this now... He opened the door, but there was no one there. Baffled, he walked outside and found Jean in the garden under the plum tree, looking up. My father called out to him, annoyed.

"What's that you're doing there, boy?"

"Nothing," Jean whispered. He didn't make a sound, he didn't move. "I'm trying to imitate a great tit. I think it answered! I think it's closer! There!"

"Answered, nothing. That little beast won't be talking back, you're singing far too high."

My father strode away confidently. I'd been standing behind him all along, and I decided to keep watching Jean discreetly. His silent silhouette was a shadow against the setting sun. I caught a glimpse of him reaching out with all his might to the little bird, trying to call out to it in the hope of starting a conversation. Thinking I'd only heard the tit's disyllabic territorial song, I had turned back to the front door, when suddenly there was a new sound. The familiar two notes of the male tit turned into a three-note ritornello. The male's song was more intense and much more powerful. I could clearly make out two birds. I went back and realized that Jean was singing with the male, who flitted from branch to branch until he landed a few inches above his head. There was nothing separating the child and the bird. Despite their obvious extreme differences, an invisible thread connected them. I was so taken in that I forgot to look for the second tit, but I soon realized that the second bird was Jean. He was whistling with a technique I didn't know—hidden, imperceptible between his lips, he murmured to the bird in a whispered whistle, mimicking passerines, almost inaudible to humans. Jean had left human language entirely to enter another language, another world.

He never showed us that technique when he came over, maybe because he was afraid he would mess up, or maybe he was ashamed: that kind of whistle didn't have the same strength as the songs of Baie de Somme birds, mostly marine birds like waders and ducks, which have a powerful call, matching their incredible capacity for movement. During storms or on foggy nights, they have to find their way around by ear, and so they've developed the ability to communicate over many hundreds of yards, or even miles, all the more so when they're surrounded by disruptive elements—wind, rain, crashing waves… Yet with that calm, fine whistle, Jean was restoring the nobility of garden songs, and humbly taking a back seat to the birds themselves.

I retraced my steps, careful not to be spotted, and decided to walk around the house. At the corner, I came face to face with Jean, his eyes wide and just as astonished as mine. As we stood staring at each other by the corner of my house, a kind of mutual challenge and respect passed between us. Jean's gaze ran through me, and I held it, unblinking. When two birds meet, there's not a word; everything is said in the weight of silence. Suddenly, a high-pitched "Johnnyyyyyy!" rang out through the street. My mother was worried that I wasn't back yet at that hour, and was calling me home. Jean followed, and another bird lesson began in the kitchen. I headed upstairs and resumed my post, perched two steps up from the landing.

That day's lesson was very specific, and sung. Jean had already managed to stabilize the notes in the whistle that just the day before had been so tenuous. He must have been practicing day and night. My father explained that, to do the *ch'corlus*—the curlew—you first had to get the trill at the beginning of the song, a sort of drumroll the bird voiced before the long melody. The roll is like a rolled *r*, produced in the back of the throat, and the technique combines the guttural *r* with a sustained, continuous whistled note. But as soon as that tickle came up at the back of the throat, Jean's whistle would disintegrate, his natural child's voice reappearing. I could hear something like a baby's guttural cooing, and burst out laughing in my hiding place. But Jean didn't give up, and kept trying, fifteen more times, thirty, fifty. My usually patient mother let out a long sigh, and my father suggested moving on.

To introduce the curlew, my father used the Picard name for the bird—the key, he said, to a perfect imitation. The word *corlus* comes from the main theme of the bird's call, beginning onomatopoeically with a few seconds of that rolled *r* followed by the *cur-loo—crrr cur-loo cur-loo ur-loo ur-loo ur-loo*—all that while whistling the right tone. While Jean practiced, only managing to get the right note every other time or so, my father told him about the great migratory bird imitators who can reproduce the song of the curlew with their fingers, using a very rare and resonant technique that can hypnotize any northern bird.

I don't know whether that was true or not, but his words stayed with me. A whistle that could hypnotize birds! To think that I might be able to become a kind of bird whisperer, talking and interacting with nature from an apple tree or a hilltop, seemed like an impossible dream. I'd always been surrounded by birdsong, and the rudiments of bird calling had been instilled in me early on. Curlews, redshanks, greenshanks, chickadees, song thrushes, blackbirds, and other birds were part of my musical landscape. I gleaned little lessons almost every day, from my father, my grandparents, and my uncles, who were all very close to birds: "Look at the chickadee feeding her young. Can you hear her?" Or, on vacation with my grandparents, I'd see my uncle come home from work imitating a common blackbird in the yard. "There's my own little blackbird singing back to me," he'd say proudly. What Jean found with my father was a slice of my world, the world I'd been immersed in since birth. Yet going so far as to conceive of a common language between humans and birds seemed unthinkable to me, almost transgressive, as if only madness or an alchemist's gift might bestow upon those imitators, those bird-men, the ability to step from one world to the other.

Jean was still stuck on the curlew's rolled *r*, but he got the singing part right, and had just about mastered it. *Cur-loo cur-loo cur-loo!* With a burst of enthusiasm, he tried out a higher note, a shrill, assertive *cur-LEEEE.* My father leaped out of his chair.

"Never, ever, ever do that!"

Jean was stunned. He'd heard it before, he replied,

and that was how the Peterson's *Birds of Britain and Europe* described the call, which after all was what had given the bird its name in French, *courlis*, just as the cuckoo's name is linked to its song. My father's eyes were steady, and he replied, in crystal-clear French this time, without any dialect: "Your *courlis* will run away when you call like that. And that's not what's written in your book; what you're doing is an alarm call. If you want to bring the bird back and get it to answer, you have to do everything in *ooo*, never in *eee*. That's why in Picard we say *ch'corlus*! *Ch'corluus, corluuuuuuu*! *Ooo.* Not *eee.*"

Jean frowned a little, dropping his head. For the bird-call competition in Abbeville, he replied, you had to be able to imitate everything faultlessly, including alarm calls, contact calls, territorial defense calls, and mating calls, and both males and females. As that evening's lesson drew to a close, everyone went back to their own activities. My father remained convinced that the *ch'corlus* lived up to its name and call, while Jean stubbornly stood by his *courlis*. Meanwhile, I dreamed of divine mimicry, a link between humans and birds.

Competition

THE CARD WE RECEIVED read, *Hôtel de France, 6:30 p.m.*

The hotel was huge, the room tiny and crowded. I thought I could hear a teal duck: the contestants had arrived. All men, most of them unaccompanied. I sat down, and everyone stared at me for a few seconds. One guy did a common greenshank call for his seatmate, and another, a wigeon. It was incredible: I was both awestruck and terrified.

There was a father and son, a little older than me, fishermen from Le Crotoy, so burned already by the April sun that it was hard to fathom all the hours they spent in the bay, living with the birds. They spoke quickly in Picard, and I couldn't understand a word they were saying; all the bird names were in Picard. I didn't have that in my books—the *pnards*, the *ringands*, the *tarneros*, the Picard words for pintails, shelducks, and terns, probably common terns, judging from their imitations.

Next to me was a man who didn't speak, and who was as pale as the mortician in *Lucky Luke*. He had a pointed nose, and a sharp gaze… Did he look like a bird of prey? No, more like a crow—a crow, watching. When his wife came over to talk to him, however, he got angry and turned into a jackdaw.

And then Zorro came in, letting the door slam behind him. He stared at us with opal-blue eyes. The tide was turning: this year, he announced, he was going to win.

"He was the last one we were waiting for," the two organizers said. They went on to remind us of the rules: there were three rounds onstage, with each contestant given one minute per round for their chosen imitations. We drew lots: nineteen contestants, nineteen little numbered slips of paper. I got five. Zorro got one. He was up. His eyes went dark; opening the bird ball had not been part of his plan.

The contestants had to declare their choice of birds, and the order. If I started off with a gull, I would make a great impression, but I'd never be able to match that. What if I wrapped up with the gull? A strong finish… though that was risky if I was tired. Second, maybe. Yes, that was it: curlew, herring gull, great tit.

As soon as I named my second choice—"herring gull"—every head in the room turned to me. Zorro, in a hideous, hoarse, gull-like cry, shouted, "*Y a pu d'pichon, y a pu d'pichon!*" There's no more fish: a sailor's joke. The audience laughed. I did not.

My sharp-eyed seatmate had drawn number nineteen; he was happy. "Birds?" Sandrine, one of the organizers, asked.

"Crow, jackdaw, rook." I was elated! I had seen the birds within the man. He became suddenly talkative: last year, he told me, he'd picked the first spot, and when he launched into the jackdaw the whole room laughed, which must have swayed the jury. This time, by going last, he was sure to rank higher.

After we had picked out lots, the contestants gathered for dinner at the hotel. The atmosphere during the meal was somewhat awkward. The contestants were all trying to get each other to talk, with the ultimate goal of getting them to mimic one of the birds they would be doing up onstage. Depending on how good these demonstrations were, the butterflies in the other contestants' stomachs would flutter harder or calm down. The clatter of forks slowed the better and more accurate the bird calls got. Some people hardly touched their plates. Zorro tried his luck, asking me to do my gull, but I refused. One of the guys from Le Crotoy saved me with a whistling gull, a fledgling cry. Very nice, thank you.

We all headed for the Ponthieu movie theater, where a big crowd had gathered in front of the main entrance. The contestants went through a small door behind the stage: the bird door.

There were two parts to the festival evening: first, excerpts from the competition films and photos, followed by the bird-call competition. Then there would be an intermission, and finally the winning film would be announced and screened.

A journalist, Denis Cheissoux, was emceeing. In the dressing rooms, we were introduced to the jury: someone from the Fédération des chasseurs, a representative of the Baie de Somme hunters' association, the director of the Maison de l'oiseau aviary, two ninth-grade students, two big names in wildlife films, and only one ornithologist, Philippe Carruette, from the Marquenterre wildlife reserve. The judges shook hands with some of the contestants. They knew each other. Was this thing rigged?

I looked out at the audience through a small gap in the heavy tobacco-colored curtain. The fabric smelled like dust. Eight hundred and fifty people would be there, in front of me, in just a few minutes. The emcee asked the audience to hold their applause during the imitations to avoid influencing the jury.

The competition began. Zorro dove right in, and, immediately defying the emcee's instructions, the crowd cheered. His song thrush was almost better than the bird itself. He whistled powerfully—the sound wasn't perfect, but everyone was amazed at the intensity. He got the repetitions right. The calls he'd chosen weren't the hardest, but his whistle was unbeatable.

The next contestant did a wigeon, a guy did an oystercatcher, and then, just my luck, another did a curlew. Right before me, a fellow from Saint-Valery-sur-Somme, a bay boy, was doing the curlew, and quite well too. Without realizing it, he gave me my opening note. I ran onto the stage as if we were in a relay race. All I could think of was that note, picking it up, not dropping it, matching his speed, his breath. I was up… I was whistling too fast, I knew it, but I had the note, the trill, and the *ooo* at the end: *cur-loo… shrpf.* Whoops; too close to the mic. I ran out of air, and stopped. The room was quiet; it felt like a long time. There was some applause. I didn't look at the crowd, and went off.

Backstage, I saw my friend the jackdaw, who was so nervous that his lips were twitching and his face was bright red. Like the birds in Jean Giono's *To the Slaughterhouse*, the crows seemed to be pecking at him from the inside. We were on number seventeen; his turn was coming up.

The audience was laughing uproariously. One of the contestants appeared to be stuck. He was doing the oystercatcher, but somehow only the contact call would come out; he must have been on his third minute of *k-beeep, k-beeep, k-beeep!* The man was as lost as a baby bird, a young oystercatcher still with its white half collar, wandering in the perilous late-summer fog rolling in off the sea. The bird was lost, the man was lost, and that lonely cry rang out hoping for an answer that might cut short its pained flight. The audience started clapping, maybe to put an end to the ordeal, but the man carried painfully on. Finally, the emcee came to his rescue, while the audience laughed and laughed.

Sir Jackdaw came on, the last one up. He was mimicking the carrion crow. As soon as he grabbed the microphone and named the bird, the audience broke out laughing. Was it something about his posture? That characteristic stoop? Memories of the previous year? I watched him from the wings. He let out a *caw caw caw caw*, the typical territorial croak. Not bad. It lacked some of the bird's metallic resonance. He followed up with the *craw-craw* when two birds see each other, but botched the higher-pitched, flutelike call, hitting it far too low. And he went all out on the cantankerous scolding of a brazen attack on a hawk or some other bird of prey that might have had the misfortune of flapping into crow territory. On the whole, it showed the range of the carrion crow's register nicely, albeit without the short-distance contact calls, and especially the hardest call, of a male crow singing to his partner: he tilts the tip of his beak down onto his chest, closes his eyes, and produces a fluttery little burp, his personalized sound signature, before opening his eyes again.

It was time for the second round. I was up before I knew it. I faced the mic; I could see the audience's eyes riveted on me. Be careful with the microphone, I reminded myself: it was a godsend for more muted calls, but a hindrance for anything too powerful. Several contestants touched or hit it, which instantaneously wrecks any vocal quality. Darn! I realized that some other contestant had raised the mic, and I couldn't reach it. Too bad; I drew myself up, held my head out like a bird trying to nudge the highest branch with its beak, and, in my child's voice, I gave a first *keow*, the contact call of the herring gull, not too loud.

The room fell silent. I tried a second, more intense *keow*. Time stood still. A third, a fourth, the rhythm closer and closer. The audience shuddered—it was hard to describe, a kind of harmony of emotion that gave me wings. I followed up with another contact call, and the higher-pitched response I imagined was a female. My body bent of its own accord, and my arms opened. Without any effort, the bursts of sound ricocheted from my body.

Keow, keow, huoh-huoh-huoh…

I didn't even need a microphone. I was a gull, I was flying. I glided, my arms wide open, and the edge of the stage became a cliff. But I ran out of breath, I forgot to inhale, and I had to stop.

The applause was deafening. The audience roared, 850 spectators in unison. I could see the judges in the front row smiling. Some scratched on their note cards, and others were clapping. From that moment on, I was the official gull of the bird festival.

I left the stage. I regretted not having included the chicks' thin, piping *klee-ew* begging cries, and the adults' warning calls, those cackles the parents repeat when they sense a threat, which makes the babies freeze in the nest—like a game of red light, green light—just long enough for the danger to pass.

Backstage, the other contestants looked at me oddly.

Some gave me a thumbs-up, to which I responded with a small smile. Even the jackdaw seemed to have overcome his struggle against the birds of darkness. He smiled at me and came over, whispering something I didn't understand. Something nice, surely. The dark cloud hanging over me had vanished. Only Zorro was still glowering at me. I looked at my feet.

Everything went quickly after that. On my third turn, I bungled the song of a great tit. My whistling technique was impossible if I wasn't totally relaxed. At the slightest contraction, my triple embouchure—my teeth, lips, and tongue—fell apart, and the sibilation collapsed into an uncontrolled hiss. Though to be fair, given that jury, I don't think a great tit, no matter how accurately interpreted, could really have scored very well. In the future, if the jury was the same, I would have to work on the typical species of the Baie de Somme. The common greenshank or the redshank. Not the wood sandpiper. A wigeon or pintail, but not the mute swan.

The good Sir Jackdaw finished with the jackdaw, as usual, and the audience laughed congenially again. Then it was intermission. I was supposed to meet my parents in the auditorium, but I couldn't even get out to go anywhere. Everyone jumped on me, like a colony of gannets on a herring shoal. "How old are you?" "You're from Arrest, aren't you?" "What's the difference between you and a gull?" People stared and stared. When I finally reached my parents, they were surrounded too. They didn't quite understand what was going on, but it was clear I'd made an impact.

After intermission, the previous year's winner, Hélios, who was also on the jury, performed the cabaret act he staged in Paris. He came on dressed as Robin Hood, with a green felt hat adorned with long pheasant feathers. He whistled beautifully. His act was a tribute to springtime. He mimicked several little birds—imaginary birds he brought to life with his fingertips and seemed to release, as if he were juggling them. He was no mere imitator: part Papageno, part Robin Hood, the man named Hélios drew us into his tale. At the end, the sun began to rise, and he cuckooed up the dawn. It had all been just a dream... That night, I understood that birdsong tells a story. That lesson stayed with me forever.

At last... drum roll... "The jury has deliberated," Denis Cheissoux announced, climbing up to the podium. Third place went to the fisherman from Le Crotoy, who had done a magnificent curlew. He had also tried a herring gull, but this obviously wasn't the right night for it. And second place went to... me! I was the second-best bird caller out of nineteen contestants, at only eleven years old. I also won the under-sixteen prize, and the tidy sum of three thousand francs. I couldn't believe it.

First place went to Zorro, who won a trip for two to Djoudj National Bird Sanctuary in Senegal. He grabbed the emcee's microphone and hogged center stage for a half hour. Like a thrush gorging on overripe grapes, he embarked upon an exceptional one-man show, imitating every bird in the bay. He sometimes confused the call of the lesser scaup with the greater, but it didn't matter. He reprised his gull scene—*pu d'pichon*, no fish—with the bay silting in, as he'd performed for us earlier. He also did a whole string of more or less raspy calls, pointing out that the song thrush essentially says, *tcho cul tcho cul tcho cul-tchotte bite tchotte bite tchotte bite*, a funny, bawdy Picard play on nether regions and bodily functions. The audience hooted, though my parents certainly weren't laughing. He wrapped up with his imitation of the little Baie de Somme train and the Noyelles stationmaster.

Eventually a less-than-stellar ranking got the better of Zorro and he declared himself out of the competition for life, to keep the lore alive, but for at least ten years, Zorro was a regular at the bird festival. No matter how he ranked, the locals waited with bated breath for the annual performance by our very own Baie de Somme Zorro. Some years were better than others, depending on how long the intermission was, and how much he had wet his whistle. It was always worth watching the audience for reactions from town officials who hadn't been warned about his antics… especially one year when Zorro had lost his driver's license and came up to the prefect to try to get it back by doing an imitation of a chicken. That is what bird calling is all about—a bridge between Hélios's poetic imagination on the one hand, and, on the other, real-life tales from the wilderness. In mere seconds, birdsongs can make us laugh and cry. That evening, I met my first two masters.

When I went to see Monsieur Rasse the day after the competition to thank him, the *Courrier picard* was sitting on the table. He couldn't get over the three thousand francs. "I told you Zorro was good. But he's never really going to go to Africa."

Child Prodigy

WE OFTEN ATE SUNDAY DINNER with my grandparents—the adults would take their aperitif in the living room with the TV blaring, talking about the latest storm, a late frost after which it was safe to start planting, or about some recent death in the family, a great-uncle or a distant family acquaintance. They shouted back and forth, talking over each other, while the Antenne 2 news chirped its familiar opening credits. Everyone watched distractedly, waiting for the same thing: the weather, by Nathalie Rihouet. Now and again, a significant current event would catch their attention: I remember once my grandmother glancing at footage of a wall that had come down, the sign of renewal between two blocs. The images seemed symbolic somehow. And my grandfather stopped dead to catch a sports report about Serge Blanco's XV de France.

At last, the weather came on, with all the usual vagaries. There was also always a short piece on Sundays, *Le clin d'œil de la météo*, a kind of impromptu sidebar, this time on the theme of spring. That day, Nathalie Rihouet, friend to all French, uttered the name Jean Foucault. My heart skipped a beat: Jean Foucault; Jean Boucault… For a moment I thought our very own mewler was about to pop through the TV set. The regular forecast came back on, and everyone froze at the mention of anticyclone activity somewhere called the Azores, an exotic, unfamiliar word, the origin and meaning of which were unknown to all of us. The upshot was that there was no risk of overnight frost for the time being, and temperatures would remain seasonal. Rihouet gleefully pointed to a radiant sun, and then went back to the topic of that week's special feature: "Jean Foucault," she said, "*l'enfant qui parle aux oiseaux*." The boy who talks to birds.

I almost passed out. It was him! It was Jean! We'd seen him just the day before. He hadn't said anything about this, and now here he was on the screen, in our living room, inviting himself to Sunday dinner with my grandparents. There was no doubt about it: the first shot behind the announcer showed our very own Jean on an islet in the middle of a pond, arms outstretched, squawking like a gull. We watched, riveted, for a few minutes, while my little brother jumped up and down on the sofa, shouting, "Jean's on TV! Jean's on TV!" Jean was telegenic—all of eleven years old, with his big blue eyes and his astounding erudition. He spoke easily, smiling, and delivered an incredible performance with his imitation of a gull. "What talent," Rihouet concluded. "I'm speechless."

My grandparents, discovering this child prodigy for the first time, were utterly smitten. My father told them the whole story, and they were honored, they said, that our family knew someone from the TV. After my father, the students at the school in Arrest, and my family, now my grandparents had succumbed to the charms of this boy and his incredible gift for talking to birds.

After dinner, I decided to go off on my own, to the small woods a few hundred feet away from our house. It had been a weird day, and maybe all the hullabaloo had worked its magic on me too, because for some reason I slipped two fingers into my mouth, my middle finger and ring finger, leaving a space between them to allow air to escape. My tongue slipped into place under my fingertips. Everything fit together naturally. Within a second, I managed a low vibration and a sound, a note. There was lots of breath warbling through it, but there it was, and with it the promise of more to come. In that little grove I escaped to with my fears—of seeing the pillars of my childhood gone, of another little bird encroaching in my nest, in the warmth of my mother's breast—I made a plan; I could see a new path.

From that day on, every day I would come practice under the branches of the tall oak trees. I would work hard on my whistle, to make it so good it would even be able to win back my father's love.

Two Boys

JEAN LOOKED HUMBLE when he showed up on our doorstep the next day. He had come to ask my father's advice on imitating the male whistler. Right away, I could sense that my family's behavior changed; they became busier, more attentive, like in a bad stage play. *Thank you, Jean*, they bustled. *Yes, Jean!* Every sentence seemed to be punctuated by his name. It surprised me what an impact the TV spot had had on my family. Jean hadn't changed a bit, but the second he showed up on the screen, everyone else was transformed. It was strange to see my parents passing around fruit juice and madeleines at seven o'clock at night. The family gathered in the kitchen, staring at Jean. But even my father's interrogation—"Well…?"—as he waited for juicy details about the TV show didn't change a thing. Jean's reply was crystal clear: "I want to imitate the male whistler."

I was struck by his honesty. His soul was untainted by the siren song of fame. I hadn't budged from my perch at the top of the staircase. I smiled. The wigeon swept away any talk of television. Imitating a wigeon was so important, so essential to Jean that his stint with Nathalie Rihouet seemed insignificant. That was when I truly started to understand what a strange bird Jean was. I inched down the stairs a bit, listening to the lesson about the sublime new duck. The call is a clear, chaste whistle, a sonorous, high-pitched *oo-oo-eee-oo*, a song that resounds from the north during the great winter migrations. My father focused on the note that should be held at the end of the phrase, but Jean still couldn't whistle very well. His mimicry was thin, like a singer with an unstable vibrato, and the high note slipped away in the excess of air.

Still hoping for gossip about Nathalie Rihouet, my father prodded again. "We're waiting..."

Jean smiled: "I was second!"

My father didn't understand, and, in a lovely northern burr, retorted, "Huh?"

"I got second place," Jean repeated, "but I still came first." The conversation was surreal. I shuffled closer.

Jean explained that he had competed in the famous Abbeville bird-call imitation competition, at the Ponthieu movie theater in front of 850 people, and that he finished second overall. But, he pointed out, he won first prize in the under-sixteen group, a category the organizers created to spare the egos of the adult contestants. In light of this child prodigy arriving on the scene, the festival had made sure to award first prize to an adult, to avoid offending anyone. Jean was pleased with his prize, and with the three-thousand-franc purse. He wanted to buy a new scope for bird-watching, he explained. You could see how happy he was on his face, in his body, and it was contagious. My father was bewildered. Why would a child who had just won three thousand francs continue to devote himself body and soul to his passion, like a scientist who never tires of the marvel of discovery? Despite my parents' best efforts, Jean kept mum about his big-deal TV spot.

Two weeks later, my mother went to the mailbox and pulled out the village newspaper, *ch'Bidayen*, which usually reported mostly on pretty gardens, local births, the fire department, and roadwork. But this issue had a full page about the achievements of the pharmacist's son.

The mail also included a strange envelope bearing the logo of the Abbeville chamber of commerce. My mother didn't dare open it, and laid it on my father's plate like she did all important documents; he would be home for lunch around twelve thirty.

I was back in my grove that Wednesday morning, out of sight. My wigeon was getting stronger, intense and resonant. After the last lesson I had worked to fine-tune the high-pitched ending of the distinctive *oo-oo-eee-oo* phrase. My call sounded exactly like the redheaded male parading around for his mate. By now I had mastered quite a few birds. My father and uncles whistled bird calls on the way home from work the way others might sing along to the chorus of a song, and wader, curlew, and redshank calls came easily to me.

However, unlike Jean, I had mastered a whistling technique using my fingers, which no one else used, and which was incredibly powerful, like a singer in full voice. The birds my father and Jean worked on in their lessons were simple, often with only one or two tones and without melodic lines or long declamatory phrases. Songbirds, those great garden virtuosos, have highly developed whistling organs, and the longer I spent alone in the woods, the more I left behind seabirds to flirt with the denizens of the forest. Their range requires precision, and, beyond the notes, the texture of the sound can vary, with trills, diphonic progressions, or a more pronounced attack. Just when I thought I'd gotten the hang of a thrush or warbler, I immediately realized that my melody still lacked grist, thickness, or clarity. Yet alone in the trees, I felt I was getting closer to the passerines' prowess.

My secret safely concealed, I headed back home for lunch, listening for my father's 2CV. I ran behind his car when it pulled in. No one was allowed to come to the table after him; we would get in trouble if we wasted the time he had for lunch. Everyone had to eat at the same time, and the same thing. That day, it was pigeon with peas and bacon in the big pressure cooker. I'd woken that morning to the acrid smell of burning feathers, a sign my mother was preparing this dish, which, like every day, would be devoured by our five hungry mouths without a thought for the work that had gone into it.

My father asked what was for lunch, as he always did, and spotted the envelope sitting on his plate. It wasn't a bill, or one of my report cards. My father steeled himself as he picked up the envelope bearing the official-looking stamp of the Abbeville chamber of commerce. "What's this?" His reaction hinted that he expected the worst; surely this meant an account was due, or that he'd overlooked some compulsory declaration.

He slit open the envelope using his knife as a letter opener and pulled out three similar cards, addressed to the three men in the family, my father, my brother, and myself: *Application to enter the Bird Festival birdsong imitation competition*. Our first and last names and addresses had already been filled in, and, at the bottom, three empty spaces awaited our reply: *Bird 1, Bird 2, Bird 3*. My father was astonished: Who could have come up with this information? Who could possibly think that our entire family were bird callers? It was clear to me who'd signed us up, though my father chose to believe that he was such a fine whistler that his reputation had reached across the whole Baie de Somme.

I later found out that it was indeed Jean who had entered us in the festival competition. If he had been able to nominate our dog Bobby, he probably would have. It was a weird ideal, born of a beautiful dream of having the community of Arrest reunited onstage at the Ponthieu theater and crowned on closing night.

But it was not to be. In a gesture of pride and contempt, my father immediately tore up his registration form: "Nobody signs me up without my permission." My brother was too young. That left me. The sheet bearing my name was laid away carefully in a drawer, somehow kept safe by my mother. I thought I'd kept my secret, and I still don't know how Jean could have known that I was hiding in the woods, gradually turning into a bird-child, undergoing a metamorphosis away from prying eyes. And I have no idea how my registration form wasn't thrown away or lost like one of the many driver's licenses, ID cards, or keys my family made a habit of losing.

On his next visit, Jean didn't say anything about the registration forms, and neither did my parents, as if that fateful day's mail were subject to a code of silence. There was no lesson that night, and Jean, my father, and I just sat out listening to a blackbird singing in the cherry tree in the garden. We had the nicest cherry tree in the whole village, a Bigarreau graft. The blackbird was happy, the cherries were dark red, and I stuffed myself with the fleshy, sweet pearls as the black-cloaked bird easily put all of its winged friends to shame: the blackbird was the master, and all other songbirds mere apprentices. Blackbirds are the most beautiful singers; my father and Jean thought so too, though I think Jean was enjoying the cherries even more. My father, meanwhile, had stopped eating, listening intently to the blackbird's drawn-out warbles. After some discussion, we agreed: it was impossible to capture the blackbird's song exactly, it was so artful and complex, with the theme changing each time to return always to the same phrase, like a cycle in a Bach variation.

Time no longer mattered, for the blackbird and for us. It dissolved beneath the song's technical mastery, and we spent what seemed like forever out there, just listening, in an idyllic landscape, with the cherry tree, the evening dampness, clutches of beetles dancing around the birches. Night fell slowly over the countryside. Suddenly, with a dissonant, percussive crash, the blackbird flew off its place. Its flight cry is worthy of a composer like Pierre Boulez—those soaring, percussive works—and the change from the melodious tune we'd heard was so brutal, so radical that it seemed like a different creature entirely.

The common blackbird has another typical cry, usually heard at dusk. All the blackbirds in a given area will give a few dry, firm notes, like a choir, a reminder that it's time for bed, for them as well as for us.

Son of the Wind

AFTER A FEW WEEKS of unsuccessful attempts in my grove, I was still sadly earthbound, far from the warblers. It was high summer, and their songs had been replaced by a chattering of starlings that taunted me all day long. By the end of July, my refuge had become a place of mockery, a communal dormitory for the starlings, with their shrill, repetitive cries.

My heart wasn't in it anymore, and the enthusiasm and hope of the early days had given way to wistfulness. I was feeling pensive. I let my gaze wander over the huge white sheep sailing by above me. My eyes swept across the sky, across the fields. In the ripening wheat, a skylark chirped tirelessly, rising out above the sheaves, but though I could hear the male's telltale song, I soon lost sight of him in the enormity of the sky. The lark's song always seems so eternal, so long for such a small bird: they don't even clock in at two ounces, but they can sing for ten minutes without stopping. The high-pitched melody goes on forever, the lulling sweetness of summer.

As I scanned the horizon for the dark spot of the bird, I spotted instead a translucent diamond shape maybe a quarter mile away, balanced out by a long, fluttering tail, and bobbing up and down like a kestrel over a field mouse. The diamond was about as high up as a bird of prey at typical hovering altitude. I was moving slowly toward this strange bird when a gust of wind swept along and it dove down to earth, straight into the tall wheat. About a hundred feet over, I noticed a child, with a cap on his head.

"Go right," he shouted. "Go right, a little bit farther, go, go, I'm telling you, it's out ahead!"

The wind was picking up even stronger, and his words carried.

"Go! Hurry up, it's windy out there!"

I obeyed.

I ran, oblivious to the stalks whipping my legs, not even thinking that my pencil case and my sketchbooks might fall out of my bag. There, on a carpet of bright red poppies, was a kite, like a fallen soldier in the Great War.

"Take it and step back, go on, quick," the boy with the cap shouted. "It's windy, can't you feel it? The weather's turning, it's going to blow, hard!"

I rushed over, grabbed the kite, and stepped back. The string went taut. "Stop," he yelled. "Face me, and watch out for the tail!"

I did as I was told.

"A little to the left," he went on. "Yes! Upwind! That's it!"

The kite seemed completely archaic; I could feel the weight of the hours spent trying to fashion the flying machine.

"Don't toss it," the boy called, "just let it loose." A warm current lifted the square of fabric upward, straight toward the sky, which had grown dark. All at once I understood about letting go in the face of such power.

Without taking my eyes off the kite, I staggered forward through the wheat toward the odd boy, who looked like he was a little older than me. He was tall and broad-shouldered, and I thought it might be Rémi, the village game warden's oldest son. He was holding not a reel or one of those spools kite-fliers have, but a huge wooden crank made by the village cabinetmaker—I recognized my friend Gilbert's work; he'd made me chickadee nesting boxes. The crank was so big and so heavy that Rémi could lay it on the ground without the kite lifting it away.

The crank was wound with heavy-duty fishing line. "Does your father fish?" Rémi asked. He did, I said, or he used to a long time ago, with my grandfather, in the Somme canal. His greatest catch was a sea trout, a story he'd told at least twenty times on New Year's Eve.

"Does he still have any line?" Rémi wanted to know.

"Why?"

He was collecting twenty-pound line from fishers in the area, he explained. I knew that a spool of fishing line could be quite long, fifty yards or more, and there were lots of fishers in the village. Judging from how much line was wound onto the enormous crank handle, Rémi's negotiations must have been extensive. Miles of different lines, tied together, as Rémi told me, with ring bend knots, unspooled before my eyes.

As I listened to his adventures, I noticed that Rémi's fingers were skinned raw, especially his index, the skin all cut up. The fishing line was slicing into his knuckles, but he hardly seemed to notice, staring at his kite as it flew straight up. I looked up too, searching for the diamond shape, but it was as though there was nothing between the boy's fingers and the sky, as if the kite had disappeared completely, sucked up into the firmament; as if the thread were stitched into the heart of the clouds. Rémi was holding the heavens in his hands. With his wooden spool, he was fishing in the sky, a cloud fisher. The line was curved, and he turned to me, his eyes full of wonder: "Here, try it, you'll see, there's no tension at all."

After tugging so hard it almost severed Rémi's fingers, the line had gone soft and loose, as if the kite had finally found peace, free at last high above the clouds. I thought of birds, of a flock of cranes or even geese, and I felt closer to them. Rémi stood there, his face turned to the sky, and I thanked him again: I had felt the wind because of him, I understood it now.

Rémi was always a dreamer. When he grew up, he enlisted in the air force, but it was a bad time: the war in Kosovo. He was twenty years old. I think he probably tried to fly too high, up in the clouds.

The Devil and the Deep Blue Sea

RESCUE STORIES USUALLY END BADLY. Maybe a baby bird fell out of its nest onto the sidewalk, or you catch the cat toying with a little fluff ball. Quick, you grab the bird and tuck it into a cloth-lined cardboard box.

The little marvel is settled in. At first, it's scared, but gradually the fledgling gets used to you and no longer startles at your every move. After a few hours, it stretches out its tiny featherless neck and pivots that little round head tufted with down, turning two big, half-closed eyes on you. The broad yellow beak is open wide, the gape flange soft and bright. How can you resist? From the first moment the baby bird looks at you like that, how can you not bring it the meal it's been waiting for? The little cries only exacerbate your desire to soothe or satisfy the little creature.

I've always been fascinated by hunger cries, whether from children or chicks. The more unbearable they are to listen to, the more parents rush to meet that need, hoping that the baby will be quiet for a second. By a crib or in a nest, parents are in the same kind of trance.

And when baby birds squawk at feeding time, they're so loud that predators can spot the nest easily. They have to be quieted, and fast, lest the little family be spotted.

Depending on the species, chunks of hard-boiled egg, mealworms, a couple of earthworms, or pieces of chopped-up steak will tide them over, along with a few drops of water dribbled out through a straw. After a few days, the little glutton will have doubled in size. Its droppings will start smelling pretty foul. At first, in the nest, the parents remove the fecal sac, like a full diaper. The sac is a membrane of soft, transparent mucus, which keeps the parents' beaks from getting dirty.

Mama blackbird would bring a few worms and leave with a fecal sac. She never leaves the nest empty-beaked! The parents take turns. Miraculously, only one baby at a time poops out a fecal sac. The parents dispose of the droppings at some distance to avoid giving away the nest's location.

As the brood matures, around the tenth day, the young birds stop producing fecal sacs. Now, with a little flip, a rocking motion, and an abdominal contraction, they eject their droppings overboard, launching them up to three feet away.

Meanwhile, in the makeshift cardboard nest, down begins to turn to feathers. The nestling can stand on its own. And that smell... It might be time to change that cloth.

The baby bird grows quickly, and it never stops begging. It learns to recognize its savior by voice or gait. The bird knows when you're there to feed it, and holds its beak wide open. But if someone else comes near, it'll cower at the back of the box warily—a matter of survival, though now and again some reckless bird might squawk for a meal from even the cat!

I had many residents over the years: the blackbird with an injured head the little girl next door brought over when I was seven; a speckled flycatcher I found frozen under a hedge when I was ten; the thrush trapped in strawberry netting; a house sparrow that fell from a roof on the farm; a scrawny turtledove; some goslings... All of them I took into our house or even into my bedroom, and they in turn opened up their world to me, offering up bird instincts and their share of wildness. In exchange, I gave them what I could scrounge of crackers or bread in our cupboards, or, better yet, some flies I gathered after long sessions with a swatter by the window, and the numerous earthworms retrieved under boards I'd laid out in the garden for the purpose.

I gave them everything my eyes could see, my ears could hear, my nose could smell. Diving headlong into the life of a bird helped me start to think like a bird.

The little cries of hunger fade as their bellies get full, and there's that priceless satisfied look on the bird's face as it closes its eyes, its suddenly heavy head drooping to the side. I watched them all like a mother watching her babies fall asleep.

Then there's a sound, a shadow, and a tired eye pops open, the little bird awake and chirping... and hungry again! It's time to go back for more worms or flies.

The fledglings get bigger, and soon they're able to fly from the chair to the table. The rug (and sometimes even whatever homework is lying around) is bombarded with pasty droppings, which the bird invariably steps in. Like Tom Thumb, you can follow their footsteps across the parquet floor and all the way to the bathroom, three little clawed digits in front and one behind.

Enough is enough! That was the decree from the family presidency. Once the birds had been saved and were grown, they had to go. Back to the wild.

Alas, all my rescues ended in tragedy. After a few hours outside, the bird would become a plaything for a cat delighted to find such an understanding hostage. The day after its release, I found the little flycatcher dead on the ground after flying into the kitchen bay window. The thrush disappeared, and the turtledove too.

My salvage operations had refloated the ship of life for a few days only, sometimes a few hours. And, I confess, I still get annoyed when someone calls me for advice on how to care for a bird they found. *You're about to have the most wonderful experience*, I'm tempted to say, *and dive into another world.* The right answer is to take it quickly to a suitable care center where the bird will never be in direct contact with humans, though these organizations are rare and underfunded, or at least be careful to make sure the bird doesn't learn to recognize you; otherwise, you'll imprint on the chick and irremediably compromise its chance at rewilding.

Becoming a bird among birds.

Metamorphosis

IN THE SUMMER, the men in my family kept busy breeding mallards, always on the lookout for the ideal choir. Every Saturday, we set off for my grandparents' place, a twenty-five-minute drive through the countryside in the family's cream-colored 2CV, heading for our family's huge, idyllic farm. My parents sat up front, with my mother at the wheel and my father basically comatose in the passenger seat next to her, exhausted from the workweek. My brother was grumpy at having to sit in the middle. As for me, I scanned the fields in the hope of spotting a buzzard or a hawk.

The car was noisy, with the roar of the engine and the windows open. Without thinking, I put my fingers to my mouth, ready to let myself be carried by the wind. I didn't know it yet, but it was time to cross the Rubicon and free myself from the ties that bound me to the little grove where I snuck off by myself to practice bird calls. My mother was concentrating on the road and my father was lost in his own world; each of us was caught up in our daydreams. The time had come for me to reveal myself.

I took a deep breath, my hair blowing in the wind. I secretly hoped that no one would actually hear me. There, tucked in the back of the car, I decided to launch into the blackbird's impossible refrain before the toughest audience in the world: my own family. I had never yet managed to get it exactly right, but a few times, hiding in the trees, I'd come close. Whether out of respect for the bird or fear of being laughed at, I hesitated. Could I really imitate that song, considered by purists to be akin with the divine?

I felt like a child about to jump off the diving board, but then backing away. The blackbird is the most complex of birdsongs, and one my father revered. Then I thought of the kite sailing easily above the clouds, and of Rémi enduring the pain of his lacerated fingers before finally reaching those heights.

I closed my eyes and breathed in as deeply as possible. I had to get through a long, continuous whistle to announce who and where I was.

The twilight song of the blackbird had been etched in me forever. Since I was born, it was my morning song and my lullaby. Each blackbird has a unique song, but they all tell the same tale: position and condition, the story of its life. The first note is a low midrange lit up by an extraordinary trill that lasts two or three seconds. This is the blackbird introducing itself, like a musician before a concert ensuring their instrument is in tune. Holding that note, which must be natural yet timbred, requires great technical skill. It is the capital letter that opens a long sentence; it sets out the singer's identity. Next are variations on a theme that hasn't quite taken shape yet but that can be discerned in a few notes; like a jazz singer covering a standard, the blackbird begins birthing a melody, unfurling it.

The blackbird is good at syntax, too. The subject fades as the theme comes in; commas allow in clauses that establish time or place, with high-pitched notes noting the bird's position or expressing some discontent, all the while maintaining the guiding line of that long sentence, its main idea. Blackbirds are the most Proustian birds: without any dead air, they can sustain impossibly long sentences, capped by that delicate end note—the bird's famous punctuation, which reminds me of grade-school spelling bees, with the teacher who always ended his dictation by bellowing, "End stoooooop!"

I still wasn't sure how I'd been able to learn the blackbird's complex structure. I was so young, yet I could imitate the unquestioned genius of the Turdidae family! Somehow that first song left a lasting mark on my approach to art, and to expressing birdsong. Like any craftsperson, the singer, having mastered the technique, can turn to more freedom and creativity.

At last, my voice rose up in the cramped 2CV. My father was startled, my brother ecstatic, and my mother immediately slammed on the brakes and pulled over to the side of the road. We piled out of the car next to an alfalfa field. They stood around me, peering at me as if I were a strange animal. I'd never have thought that a simple blackbird trill could have the power to stop a Sunday afternoon convoy. This was enormous: not even last year's chicken pox, not even my father's horrible flu had been able to stop these weekly trips. My father started pacing around me. Out of nowhere, he shouted, "Wigeon!"

A long silence ensued.

"Wigeon?" he said again, more quietly, sounding me out. Suddenly I saw in his frown what he wanted. Spurred on by the whole family staring at me, I stretched out my arms. I placed my fingers to my lips once more, and took a deep breath. *Oooo-weeee-ooooo*! Male whistler! Everyone stepped back three feet, impressed at the blast of wind that had just come out of me.

"How'd he do that?" my father exclaimed. He turned to my mother, then my brother. "Did you hear that?" Clearly they had, and they were amazed. "The *oigne*!" Wigeons where we're from are referred to as *oignes*, and they are the symbol of the Baie de Somme. To mimic an *oigne* is to identify yourself as one of us, a code for locals to recognize one another.

My father was already looking at me like I was some kind of trophy, while the rest of my family stared. With a slight smile, he started repeating the names of birds like a challenge: Oystercatcher? Redshank? Sandpiper? I complied with disarming ease. This version of my father, taking on the role of the emcee, I knew well. He put on a big show, slyly rattling off a ream of uncomplicated, two-note birds, like a warm-up before the headliner. He went on with plovers: Golden plover? Silver plover? And then back to ducks: Green-winged teal? Male or female, I dared ask, but I realized quickly I shouldn't interrupt the audition. Pintail? I knew the list of birds by heart; I'd heard it since I was a child, in family lore and at big New Year's gatherings. And I had memorized the description and details of each species during Jean's lessons.

As I think back now to those fifty-odd species, their calls I learned and performed as if it were as natural as breathing, what has stayed with me most is the drawings I made of the birds, sketched in pencil on thick sheets of paper. Some I posed on a pond, others with a mate, some in flight. Each bird had its own posture, action, and imagination. What did it eat? What other animals would they have been with? What sounds did they make? I can still feel those two gestures coming together: the act of drawing, how I'd created each bird out of thin air since I was young, and the song that brought them instantly to life.

"*Miaule*?" My father's voice pulled me out of the bird catalog. What did he mean? I was eagerly awaiting the name of the blackbird that would have enshrined me in the avian pantheon. Why was he asking for a *miaule*, now? "Gull, if you prefer," my father added tersely. I knew full well that *miaule* means "gull" in Picard, but that bird didn't whistle, it shrieked, and people in the bay area couldn't have cared less about it. I could feel my voice about to rise against my master. My answer was clear: "I can't, that's Jean's bird." The disappointment in my father's eyes was obvious. After a few seconds, he retorted that it wasn't the Boucault boy's bird. He looked straight at my mother and the rest of the family, defying anyone who would stand in the way of his newly discovered prodigy's talent. The Rasse family, he added, had always been close to birds.

And so Jean became *le fils Boucault*. He had lost his first name the second I put my fingers to my lips. I was dreading his next lesson—seeing him come beaming through the door in search of new sounds and new challenges. I worried that my father's pride would crush the innocence of a boy who'd come to think that he belonged to our clan. My father had erected the name *Boucault* like a barrier between our two families. And there I was, wanting nothing more than my father's love.

Trophy

MY FATHER WAS SITTING in a tall bar chair at the Café Chupin, opposite the Noyelles-sur-Mer train station, telling his latest Baie de Somme anecdotes to a rapt audience. The air was thick with the smoke of Maïs cigarettes, those famous blue Gitanes rolled in corn-yellow paper. The light filtered through, and about twenty people were talking and shouting over each other. Rounds were ordered and shared, tensions fading as each glass was emptied.

I found a quiet spot in the hubbub under a pinball machine, which rattled, inviting me to play, though I was more interested in the game of Crazy Eights I had going with the café owner's son. Hiding under the machine, amid much laughter, cheating, and dares, only winning mattered: the chaos above disappeared, and our cards were the most important thing in the world. As I was about to lay down my last card and finally take a hand, I heard someone calling, "Johnny!" It had to be my father, but he sounded different, not as stern. I dropped my cards under the pinball machine and wiggled out of my makeshift cave. All I could see were green mud-spattered boots and camo pants. In the space of a few card games, the bar had filled up. It was packed. I couldn't spot my father's legs in the crowd. A wild, pungent smell had taken hold in the bar, and the atmosphere had changed; it had the chaotic feel of the market.

My father waved me over. He was with a few friends I didn't know, and talked to me as if I were an adult: "Could you whistle for my buddies? Your old man is asking." His request sounded off: my father never used the word "buddy," to say nothing of referring to himself in the third person.

But the very strangeness of the situation made me do as I was asked. I stood in the boozy fumes and the smoke, took a deep breath, stuck my fingers in my mouth, and, just as I had the day before in front of my family, I performed the song of the blackbird. As soon as I whistled the first note, it was as if the fog lifted. The chatter came to a halt, and every single person in that café in Noyelles-sur-Mer was staring at me, scrutinizing me, waiting to see what came next. I had their undivided attention; I could feel them hanging on my every note. I ended with the flight call of the blackbird, as if to close my melody. I waited. A long silence settled over the crowd. I was only supposed to whistle for my father's buddies, but it's impossible to recreate the virtuosity of a blackbird without a certain amount of power. Embarrassed, I cracked a smile at my unexpected audience, and I was about to apologize to all these people for interrupting when I heard someone call out, "Magnificent!" The room broke out into thunderous applause. My father grabbed me and hugged me tight, dragging me around to the tables.

"Who's that?"

"That's your kid?"

"He does the *corlus*, too?"

"Incredible!"

All the men were asking questions and complimenting me in Picard, and slipping coins and bills into my hand. My father, faced with such success, whispered softly to me, "You'll make a living at this!"

Was his prediction temporary alcohol-induced insanity, or sudden clairvoyance? Whatever the case, it was surreal, in my child's mind, to imagine that talking to birds could become any kind of profession. I had no desire to be a bird translator or a bird interpreter or any other such invention, or even a kid going around the bars to make some cash. There was no such thing as a bird impersonator; this wasn't a career. When I got to the last table, a little old man called to my father: "Hey, *tcho* Rasse!" He was wearing a cap on his head and held an unlit, hand-rolled cigarette in his mouth. My father turned around.

"*Ches tin fiu?*" the man asked.

"Yeah, he's my kid."

"He's a good one, but not as good as my brother Zorro."

My father shuffled me away abruptly and turned back to a long, animated discussion with the old man with the unlit cigarette. I slipped away, wondering who Zorro was, and why I wasn't worthy of him. We never really spoke of that interaction again. When my father repeated the story countless times to the family, to all my aunts and uncles, it was all about the praise, and the man with the unlit cigarette was relegated to oblivion. At every family dinner and to every new acquaintance, I became the boy who talked to birds, transforming a living room instantaneously into a dense forest or an infinite stretch of seashore.

This Means War

WHEN JEAN DID COME OVER again, it felt very natural. He walked in, my father beckoned me down from my usual spot on the stairs, and, at his request, I did a curlew call. Jean didn't seem surprised. Without a hint of jealousy, he nodded and smiled, as if he already knew what I was capable of. "That's great," he replied mildly. "We've got to enter him in the competition. He's going to win! I think you have the registration form."

My father's scathing reply was not long in coming: "Of course he's registered, his mother already sent in the form."

The competition was fully eight months away; only a week had passed between my revelation in the car and Jean's return to our place, and already I was signed up for the Abbeville competition! I didn't even know what birds I wanted to imitate, and the festival asked contestants to provide a list of three. Without asking me, my father and mother had decided, selecting the curlew, redshank, and blackbird from a list I wasn't even told about.

But when Jean ventured to ask what birds I'd be imitating at the competition, my father replied curtly: "It'll be the curlew, because that's all he can do at the moment."

My father moved on to the day's lesson, which turned out to be less rich and much less precise than usual. He wasn't talking about songs or calls now, but about plumage and colors. Jean could tell that something had changed. For the first time since he'd started coming over for lessons, Jean decided to leave of his own accord. After that day, he only came back now and again, just to say hello. As soon as the door closed behind Jean, my father got on me to practice the curlew and the redshank, ending with the blackbird. I stopped midsong.

"Why did you tell him I could only do the curlew?"

"Jean can't know what you can do," my father snapped. "You whistle much better than he does, and we're going to beat him."

This meant war, then, with Johnny Rasse's blackbird pitted against Jean Boucault's herring gull. Two very different styles: the former warbled melodies from the treetops like a bard with a heavenly lyre, while the latter was Ulysses trumpeting his exploits on the long voyage home. One whistled, the other screamed. Yet both were just children mimicking birds. It was impossible to choose a favorite. One child, standing in our kitchen with his arms outstretched, turned into a seabird in the blink of an eye; beyond imitation, he was a shaman unknowing, he was gull through and through. The other, in the shadows, listened to the forest, a child raised from birth in the world of birds, who had a whistle that stopped human time, and especially his father; he was a bird-child, a link between the music of nature and the human realm.

After hours and months of rehearsal in the kitchen at Casa Rasse, my blackbird was shaping up nicely. I had it down, and I could do runs and trills more lyrical than even the bird itself. My singing charmed everyone who came over—neighbors, uncles and aunts, grandparents. My father was the conductor, directing the whole process; he was my singing teacher, pushing me to work on new sounds and explore new territory. I also added thrushes and robins to my repertoire.

Robins are great communicators. I spent hours listening to them, and talking to them. My wintry Wednesdays were dappled with the song of robins, and every time I learned a new tune I made sure to crack it out in the kitchen that evening. For my father, it was yet another chance to show me off to my aunts, uncles, and grandparents. Thanks to that little bird I also discovered the powerful high notes of garden passerines like the wren, which, with its weight-to-songpower ratio, is one of the best singers. Unfortunately, the diminutive gray bird with the stubby tail sang too fast for me; I was outclassed by the little lord of the hedges.

The robin was no less powerful. A touch fleshier than wrens, robins stand poised on those matchstick legs, and the sound from their wide-open beaks pierces the steely winter sky. Robins fire arrows of sound, like singing Cupids, and woe upon the unfortunate targets of those cute little harpoons, which is anyone who crosses into their territory. Belligerent and reckless, the robin is a fighter. In spite of their small size, robins can be incredibly ferocious, especially males against other males. After those first few opening notes, they dive into a continuous melody, like beads falling through the air.

Robins are also peerless poets, particularly adept at alexandrines. They seem to keep a kind of meter, marked by a brief caesura at the heart of their melody, breaking at the half line of the hemistich in the art of poetic recitation.

The Names of Birds

THERE ARE OVER TEN THOUSAND species of birds. Like all living things, they are assigned a genus, which is written with an initial capital letter, followed by the species in lowercase, all of it in Latin and in italics, per the binomial taxonomic system invented by the Swedish naturalist Carl Linnaeus in the eighteenth century.

"No, that can't be right! *Meleagris* means turkey!"

We were studying Latin, a text about the Romans enjoying a banquet of *meleagris*—the only word in the text that I knew how to translate, turkey. However, in spite of the image included next to the paragraph, it's unlikely that turkey would have been served at all in those days. Turkeys didn't show up on European tables until a hundred years after Spanish explorers brought them back from Mexico. We are being made to study a web of lies, I proclaimed; couldn't schoolbooks at least avoid flagrant anachronisms? In response, our teacher turned several different shades of red before going pale.

I felt like I was watching a turkey that had lost a joust with a rival: the bright red wattle deflated, fading to a silvery-blue gray, the color of fear.

At the back of the classroom, there was another man, who sat woodenly, taking notes. Our academic inspections were being held that day. Every time I opened my mouth, it was clear that our teacher, whose field was dead languages, would clearly have preferred to crawl into the mausoleum of Greek and Latin. I have no recollection of the rest of the class, but the following week, as soon as I walked in, I was greeted by an ominous request: "Boucault! Come see me at the end of the class." It didn't bode well.

But when I went to talk to him after class, our Latin teacher was downright gleeful. The inspector, he told me, had appreciated the discussion on the implausible aspects of the text and the debate that took place in class. Students had to maintain a critical outlook when approaching Latin, he said. Thanks to me, our teacher got a twenty, the highest mark on the assessment, which would help him get ahead. After that, no matter what I wrote that year, no matter what I conjugated, a pretty red "A" was scribbled at the top of my work.

I never told him, but I realized later that the reference was certainly to *Numida meleagris*, the helmeted guinea fowl, from the African savanna, which the Romans were already consuming, and not to *Meleagris gallopavo*, the wild turkey. I was too eager, and had gone straight to the genus, not the species. The following year, we had a different representative of *Homo sapiens* at the front of the class, and my papers landed on my desk marked with a rather less stellar "F" and with a chorus of *Malus, malus, malus!* I gave up latin soon after that. When it comes to ornithology, it's best to go to the Greeks.

Most Latin bird names come from latinized ancient Greek. Among my favorite Latin names are *Caprimulgus*, the nightjars, from *capri*—goat—and *mulgus*, to suckle. Not the most obvious association, for a moth hunter.

Beyond the Latin names of birds, being around them every day allows you to recognize individuals within the species. There's that recognition every year when the first male swallow returns each spring to my friend Jérôme's bathroom, sneaking in through the window to build its nest above the toilet tank, or when that chickadee with the atrophied leg or the unusual song comes back to the garden. That's your bird, you know it. Injuries, deformities, and plumage aberrations make it relatively easy to recognize individuals. Behavior, on the other hand, can change from one year to the next, depending on hormones: a dominant blackbird may find itself at the bottom of the garden hierarchy the following year because it caught some parasite or other. The quest for individualization is a necessity for ornithologists. In the old days, monks would mark swallows with small strands of colored wool to identify them when they returned the next season. In 1900, a schoolteacher named Hans Cornelius Mortensen was the first to track European starlings, *Sturnus vulgaris*, identifying them by writing the name of his village in Denmark on an aluminum ring placed on one of their legs: bird banding was born. Today, each banded bird is assigned a number, with its own complete data sheet, and measurements are tracked, as well as sex, feather condition, and body fat (reserves of which indicate available resources for migration or overwintering).

The miniaturization of satellite tags has allowed ornithologists access to a treasure trove of data: Where is that bird right now? So we can know that, on February 15, a common cuckoo, *Cuculus canorus*, coded "PJ" and banded in England, began to travel across the Congolian rainforest. Then, greedily, in one go, in early April, the bird leaped across the Sahara, the Atlas Mountains, and the Mediterranean, landing in Spain. On April 20, PJ arrived in the south of France, and three days later its burbling call sounded just a few miles from the nest in Suffolk where it was born. The record six years of monitoring by the British Trust for Ornithology produced invaluable information on the dangers that in the past fifteen years have led to the disappearance of more than half of PJ's friends, and with them the soundtrack of spring.

The Big Day

I WOKE TO THE gurgle of water filling up the tub that morning, and the smell of bubble bath. I was still lying in bed, but everyone else was already bustling around. My mother came in, warning me that I had to get myself washed up right away; I didn't even have time for breakfast. My family was dressed to the nines, too, in new T-shirts and sneakers. My father had on what was for him the equivalent of a three-piece suit: jogging pants and brand-name running shoes, probably from his soccer days. My mother was wearing earrings he'd given her for Christmas.

Today was a big day: at six o'clock, at the Hôtel de France, I had a date with the country's migratory birds. During the drive, my father was relentless, making me practice my blackbird, curlew, and redshank, so much so that my brother and sister ended up plugging their ears and making faces.

My mother kept reminding me to be polite and articulate. The host that evening, though I'd never heard of him, was none other than the great journalist Pierre Bonte. The closer we got, the more overwhelmed I was by the onslaught of advice, comments, and opinions about the evening—which, as far as my family was concerned, was the most important ever. Everyone would be there: my uncles, my grandparents, my godfather and godmother, the neighbors, my teacher, the whole village. And, of course, my brother and sister, who would be sticking their tongues out at me from the audience. We pulled into the parking lot in Abbeville at quarter to six.

The huge, luxurious hotel loomed before us, its facade shining like a mirror. I inched up to the glass, innocent as a lark. The glass distorted my reflection, making me look taller. I seemed enormous, and much more confident than I actually felt. An old lady in a wide feathered hat stepped out, looking me straight in the eye.

"Hello, young man!"

I held her gaze, though I was distracted by an inconsistency in the bouquet of golden pheasant feathers on her head: two common pheasant feathers had crept in among the others. I didn't have time to inform her, and just replied politely, "Bonjour, madame." I tried to speak firmly and clearly, and to practice my diction.

Inside, the hotel had wall-to-wall cream-colored carpet, and the reception desk, which was about thirty feet from the door, seemed impossibly high for a nine-year-old boy. Fortunately, a hostess appeared before I got to the counter.

"May I help you, monsieur?"

Being addressed formally by a woman in high heels made me grow taller still. The reception desk seemed to flatten down to nothing more than a tiny bump. I could have stepped across it.

"I'm here for the bird-call competition."

"What's your name?"

"Johnny Rasse."

My name had never sounded so loud to my ears; for once, no one was teasing me about my first name, which always drew comparisons to the rock star Johnny Hallyday.

"Please follow me, they're expecting you."

Those words, uttered by a stranger at the Hôtel de France in Abbeville, would stay with me forever. What was this child who answered to the name of Johnny Rasse doing in such a huge, fancy place?

I walked into a big conference room. A few groups of three or four people were laughing and chatting enthusiastically. Sitting in a chair in the middle of the room, one man stood out. He looked like he had stepped out of an American TV commercial: sharp haircut, teeth whiter than his shirt, tailored suit. He stared at me like a bird of prey, and I froze.

"Hello, Johnny," the eagle-man called out to me in a smooth, even voice. "I'm Monsieur Désérable, director of the Abbeville bird festival. Now that you're here, we can draw numbers." He handed me a pen adorned with a gannet's head, the festival's symbol. "You can sign at the bottom of the form."

There were contestants of all ages—teenagers, young adults, fortysomethings, and even a few older folks. The director said that someone innocent should do the draw, and it was agreed that the honor should go to the youngest. People's faces lit up, and everyone had the same name on their lips: Jean. Jean was there. I hadn't seen him. Jean emerged from the crowd and walked over to the director, who was holding out a hat. But the golden eagle turned him away.

"No, Jean, not you. Johnny Rasse is the youngest this year."

Everyone was surprised, probably at how small I was and how angelic I must have looked, but even more so at Jean greeting me familiarly. I could hear people asking him if we knew each other. I don't know what he said, but I could feel them watching me, as if I were mysterious or even scary somehow.

Before the draw, the eagle, Déserable, reminded us of the rules. We would take turns performing our chosen birds for a duration of one to two minutes. A microphone would be placed in front of us at center stage. This year's competition would be blind: no contestant would be announced by name. The jury of experts, which included animal documentarists, ornithologists, and naturalists, would be hidden at the back of the stage. We wouldn't do three calls in a row, just one per turn. The audience wasn't allowed to applaud, in order to not sway the jury; otherwise, contestants risked being disqualified. It seemed odd to stage a performance before an audience and ask them not to show any emotion at the beauty of a bird's song.

"Fauquemberg… Caubet… Norel… Leroy… Plé… Rasse… Boucault…" I would be eleventh, and Jean, fifteenth. He came over to me. I'd drawn a good number, he told me; going first was the worst.

We moved on to dinner. The menu bore the rather comical title *Retour de chasse*, back from the hunt. The tone was set: it was open season!

The other contestants were doing exclusively waterfowl: waders, oystercatchers, redshanks, and wigeons. They were almost all in their forties and dressed head to toe in khaki, with waxed jackets or wool sweaters, depending on where they were from. There was a notorious class division between Le Crotoy people and those from Saint-Valery-sur-Somme. The story was that Saint-Valery, as a port city at the mouth of the Somme canal, had had years of sumptuous trade, with ships coming to unload goods that were then barged up the canal, while the more primitive townsfolk in Le Crotoy survived on cockles, the dregs of the sea. That night, the rivalry was palpable: Which bird caller would be the worthiest representative of Saint Hubert, patron saint of hunters, and win the grail?

The long hunting tales worthy of Maupassant wrapped up as we finished our duck breast. I said nothing. Jean was watching the strange spectacle as closely as I was. Fauquemberg, whose profile was reminiscent of a male wigeon, stretching his neck and sticking his head out every time he made a sound, was sitting to my right; Plé, who had a nose pointier than a curlew's beak, sat across from me; and then there was Norel, whose redshank was so powerful I had to plug one ear to protect my fragile eardrums. Jean and I glanced at each other; we knew that none of them could compete with our favorite birds. We were shoo-ins!

A love of hunting shouldn't be the only motivation for imitating birds. Why choose only game species? They weren't usually the loveliest singers. And why only practical calls, intended to draw birds closer? The alarm call of avocets was among the most beautiful in our marshes. I couldn't imagine a hunter imitating a herring gull. They'd be a laughingstock. And what about the nightingale?

We were just children, still so innocent; maybe that was why we could grasp the intrinsic beauty of birdsong. All we wanted was to celebrate nature. We practiced and practiced every day with the sole purpose of singing like birds. There was nothing else, no ulterior motive, nothing but love.

When the meal was over, a TV journalist interviewed some of the contestants. I could tell he wasn't looking for the best imitator but the most ridiculous, the one who would waddle around like a turkey or peck like a chicken, or the one with the strongest regional accent. The fact that I was the youngest was an attraction too, but I didn't want to be interviewed. I didn't want to give away what I cherished most.

Soon it was time to leave for the theater. Our motley procession headed out, marching through the streets of Abbeville. Every time one of the contestants launched into a call, the others responded—an incredible din. The men from the French Basque Country did their doves, and the Picards their waders and ducks. Our band of merry whistlers drew attention, and onlookers crowded around, heckling us and laughing.

Suddenly, a few yards from the theater, the group stopped dead in front of a chestnut tree in the courtyard of a brick mansion. A blackbird was singing, and the whole ridiculous parade immediately fell silent. No one uttered a sound. The amazement, the admiration, was unanimous: the blackbird's song was like no other, and no one would ever be able to imitate it. As the fluttery notes dropped around us, we were brought up short for a moment, revealed for what we were—pale usurpers about to be swallowed in the depths of the Abbeville municipal theater.

The Italianate building was stately, a reminder of the town's onetime wealth. Inside, every seat was taken, the local who's who mingling with fans who had come to hear.

The curtains opened. On the stage stood a lectern, and a wide, white movie screen covered the back wall. The screen filled with a whirl of color, and a brass band played something like a theme song.

Pierre Bonte came out to greet the exuberant crowd. "Please give our contestants a warm round of applause!" I was right behind Jean, walking up a little wooden staircase in the middle of the flock of bird-men. Some of them had gone pale, and others looked terrified, their lips tight and dry—a far cry from the loquacious bunch we'd just been sitting with at dinner.

Pierre Bonte caught my eye. "I can see there are some very young people here tonight!" His voice was soft, like a children's TV show announcer.

A microphone stand had been placed in the middle of the stage. The presenter asked each contestant our name, how old we were, and where we were from. There was some silliness, and much laughter from the audience. Many of the contestants were from the Baie de Somme, the Baie d'Authie to the north, and Mont-Saint-Michel to the west. Some from the south of France, between Landes and the Pyrenees. The audience seemed enchanted by the range of accents.

After introducing the final contestant, Pierre Bonte officially declared the contest open. "From here on in," he said, "we will hear each contestant's number and the bird they are imitating. The species will also be projected on the screen behind us." The lights dimmed and the mood changed.

"Contestant number one," he called solemnly, "please step up to the microphone."

Fauquemberg, who was a stocky Baie de Somme guy, came up. He was obviously nervous but did manage a little wave to his family, who responded with whoops of encouragement. He was immediately interrupted by the emcee, who threatened to disqualify him on the spot. Chastened, Fauquemberg faced the mic and launched into his imitation of a male wigeon. He was dignified for all of ten seconds but then couldn't help but embody the duck, sticking his head far ahead of his shoulders like a jack-in-the-box. The range was right, but he had too much air, which unfortunately was amplified by the microphone. He ventured to imitate the female, slouching to illustrate her submissiveness. He switched back and forth between the female and the male, performing both sides of the courtship display, right up to the female's characteristic whine, and including a physical imitation that became increasingly elaborate as he got closer to his finale, to the delight of the uproarious audience. But the performances couldn't go over one minute, and we weren't supposed to try to influence the audience, so Pierre Bonte cut his love scene short.

The second contestant was a young man from Le Crotoy, a sixteen-year-old by the name of Johnny Marseille. He was very shy, and the geographical contradiction between his last name and his hometown led to some chuckles when he was introduced: "Johnny... Marseille? But I thought you were from Le Crotoy? What's that all about?" He was an odd kid, stepping up to the mic nonchalantly, long arms dangling by his sides. The poor guy didn't have much technique and was mostly just whistling, though under pressure he could only manage a parched trickle. It occurred to me how much courage it must have taken for him to get up in front of so many people like that, overcoming his shyness even briefly for the possibility of becoming something else.

Number three was from the Baie d'Authie. He whistled with his fingers, a powerful sound that was even louder in the mic. He did the oystercatcher, a black-and-white wader with an orange beak also referred to in France as a magpie because of the color of its plumage, producing a fairly simple disyllabic sound, a low note and a higher one, *hueep.* His imitation was accurate, but too monotone to reflect the prowess of that beautiful bird. What was more, because he placed his index and ring fingers in his mouth, with his middle finger pointing upward, he was essentially giving the audience the finger as he whistled, to another round of raucous laughter.

The next contestant was from Mont-Saint-Michel. His curlew was successful and quite loud, but the microphone was too close to his mouth and betrayed a parasitic sound from the back of his throat, as if another voice wanted to emerge, to be allowed to speak at last. Unfortunately for him, birds never produce any throat noise, since their vocal organ, the syrinx, bears no resemblance to our own. The respiratory system of birds differs from humans': a bird's body is like a balloon, and they can produce incredibly acrobatic sounds thanks to tracheal membranes as air flows through the syrinx. Few humans, however skilled, however birdlike, can ever come close.

One after the other, the contestants came up to center stage, each one with a distinctive style: some were smiling, while others seemed sad or stressed, whistling a little tune lost in their own world. One was a real character, a man with a lilting Landes accent who looked uncannily like Charles Aznavour, and who imitated a bird I didn't know, the red partridge. He placed both hands in front of his mouth, facing the microphone, like he was playing an invisible trumpet. From the very first noise, which sounded like a confident belch, the audience was laughing. He followed with a strange melody, and wrapped up by gyrating his hips. The crowd lost it. Pierre Bonte was furious, and, his face bright red, desperately tried to calm the audience.

Then came a charismatic thirtysomething man with slightly bowed legs. He was wearing leather pants and looked like a rock singer. He stared pointedly at the emcee, waving his arms, as if to ask him to stop. Pierre Bonte didn't understand, and the contestant, who was obviously forbidden to speak on pain of disqualification, repeated his maneuver, pointing at the microphone. Bonte asked him to come closer to whisper his request, and finally understood. "Ah! He doesn't want a mic. Cut the mic!" Finally a cappella, Elvis stepped forward, his legs wide apart, and launched into a full-scale imitation of the redshank, performing the wader with such force that the audience was blown away. He was an imposing presence, his body and head thrust forward and reaching out to the audience, almost beseeching. The spectators in the front rows covered their ears at the power of this ornithological rock star.

It was almost my turn; contestant number ten took the stage, an eighteen-year-old from Le Crotoy. He could have been number two's doppelganger, and sure enough, he was Johnny Marseille's brother, two years older. He stepped up to the microphone and did a pintail, a duck with the same song as the teal his brother had imitated, only deeper. Unfortunately, for a good minute no sound came out. With his head tilted to one side, the mirror image of his brother, he lulled us gently to sleep.

My turn. I was so short that a stagehand had to lower the microphone stand. The theater was dark, and I couldn't see a thing. It's quite something to imagine a thousand people looking at you without being able to see them. Strangely, I wasn't nervous at all. It seemed much harder to me to sing out in the wild, with the humidity, the wind, or the rain, than here, in this enclosed, protected space. And what judge could be more critical than the bird itself? I knew how to sing in front of an audience, I did it all the time at house parties or at the Café Chupin. I placed my index and middle fingers in my mouth, striking the best balance between stability and power, to create a sort of mouthpiece on which I delicately placed my tongue. I remembered my father's advice—"Take your time, don't go too fast"—and above all, I thought of the forest, the thick smell of humus, where the dark-suited king reigned high above the treetops. The blackbird is always waiting for an answer: a ten-second song consists of five seconds of warbling, then, for the next five, the bird waits quietly. The silence is part of the song, and it holds the open spaces between the notes, allowing them to alight on our souls.

I imagined the wind in springtime. I was at the top of a poplar tree rooted deep in the ground, singing my song. The tiny echo in the mic surprised even me. The psychic effect was instantaneous: in five seconds, the trees, the grove's rich black earth, the previous winter's leaves, and the spring breeze sprouted around me on the stage. The five seconds of silence allowed each person in the audience to enter the landscape with me, all of us huddled in awe in the same temple.

Everything became instantly clear. A silence like nothing else we'd heard since the start of the competition settled in like mist on an April morning. With each phrase of the song, the forest unfolded, becoming denser. Trees appeared, ferns waved. *Welcome to my grove, where together we can be captivated by the song of a bird that can paint a whole landscape.* I could feel that the audience was with me, and my blackbird started taking flight. Moving away from the microphone, I crossed the stage from left to right, I filled the space with sound. I improvised a chilling flight cry. I wove a whole tale for the rapt audience: they were transfixed by the distress in my voice, as if suddenly they could actually understand the bird's language. Drawing on some deep ancestral memory, we escaped together. All of us could understand a universal language beyond the barriers that separate humans and animals. Finally, I stopped. My two minutes had to be up, though Pierre Bonte, for the first time that night, had forgotten to keep time. I stepped back, and suddenly the silence was resonant: a murmur flooded the room, lapping at my feet. It was like every person there needed to share the experience.

When I went back to my seat, Jean winked at me and patted my shoulder, as if to congratulate me. Soon, it was his turn. When he stepped into the spotlight, the same murmur coursed through the theater. Jean hadn't even started his imitation yet, but the audience already recognized him. His reputation preceded him, and he'd been popular with the crowd all evening. No one knew his name, they all just called him the seagull—quite a blunder for bird lovers, since there's no such bird, technically: Jean was a herring gull.

Standing in front of the microphone, he waited a few seconds for the audience to quiet down. When he lifted his arms slightly, a shiver ran through me. From where I was, it felt like a draft had just entered the room, and I realized that before imitating his patronus, Jean was conjuring the northerly winds that carry gulls aloft at dusk. Jean reinvented himself as a gull, covered in feathers, waiting for the right gust of wind to be able to take off. Suddenly, there he was. The first cry unfurled from his belly as he spread his arms wide. The call of the herring gull pierced the air at the Abbeville theater. Each cry was an arrow straight to the soul. Then… silence. Jean hovered beside the mic, arms outstretched, and a new cry echoed like a thunderclap. He was playing with the wind, squealing and settling, arms half-open, flying toward us.

I had watched in awe when he was transformed in my kitchen, but I had never seen a transfiguration like this; Jean's metamorphosis that night seemed to involve even the elements. His cry was so true that it transported us into the violence of northern storms. Jean imitated a bird in flight, swooping back and forth in front of the microphone as if he were inviting us to follow him on his journey. It was all there—the rain, the storm, a struggle, the tumult, hope, a lull, and then at last the glimmer of sunshine that pushes through the cloudy sky and calms the rippling sea. Jean walked away, and returned to his seat. The audience leaped up, applauding. The crowd had decided to give a standing ovation to a child who could summon spirits from beyond the waves, and Pierre Bonte couldn't do a thing about it. I was moved by their fervor, and happy to see Jean happy here in his garden, his bird festival.

The competition went on, but it was more or less over at that point, and the other two rounds wouldn't change anything: that night, the sea met the land, the trees in my grove facing off against a northern storm under the Abbeville sky. That concluded the competition, Pierre Bonte announced; the jury would deliberate during the intermission. I wanted to see my parents as soon as possible, but I couldn't get to them; everyone in the audience stopped me, already declaring me the winner. Making my way along a row, I bumped into Jean. He smiled at me: everyone was saying he'd won, he told me. I didn't know who to believe. My head down, I tried to blend into the crowd, hoping to get to my family. I finally spotted my mother, who nodded to me, congratulated me, and asked me to come up to the balcony, where my father and brother had gone to have a drink.

Sitting on a bar chair, leaning against the counter, my brother was sipping a soda. My father sat next to him, nursing a glass of wine. They were surrounded by people I didn't know. I tapped my brother's arm, but a hand yanked me by the shoulder. It was my father, who was desperate to introduce me to his new friends. There I was, in the middle of a group of strangers who were staring at me and bombarding me with questions. Did I have anything in my mouth, they wanted to know. "How wonderful, bravo, how did you get started?" one lady asked. "You're amazing. Did someone teach you?" another wanted to know. "I thought I was in my own garden," someone else commended. I couldn't answer, so my father filled in: what I'd shown that evening was barely the beginning, and I was capable of imitating many more species. A real crowd started to gather around me, and I felt like the star of the intermission, until… "He's so cute," one woman chimed in. She was looking at me like I was a toy. Before she turned to walk away, she added, "But the seagull is extraordinary." They all started squawking among themselves; yes, the seagull was amazing. My father put an end to the debate, explaining that Jean had been coming by our place regularly, and that he was the one who had taught these two bird-children. He was holding court while everyone made predictions. One thing was for sure: the winner would be either Jean or me.

A shrill noise rang out: the bell signaling the end of intermission. I made my way along, with my father and brother at my side. Suddenly, an old man called out to my father: "Hey! *Tcho* Rasse!" When my father stopped to say hello, the man grabbed my hand, staring at me with eyes so clear and blue they bored a hole in my soul. "You whistle good, boy, but you're no match for my brother Zorro!" I recognized the voice, and the man. He was the old guy from Café Chupin. I shuddered. He was goading me because I had dared to challenge a god, his brother Zorro. My father awkwardly hauled me along. "Hug your mother," he told me, "and go back to your seat." I was haunted by the man's eyes, by his low, steely voice. But who was this Zorro, this legend who seemed to live a secret, masked life? I couldn't help but think of a duel, my body run through with the masked avenger's sword as he laughed ominously astride his steed Tornado. I didn't quite believe he was real, but my father seemed jittery, which was no comfort at all.

Face to Face

I WAS JOINED ONSTAGE by two new contestants: Johnny, who was two years younger than me, and a man from Baie d'Authie, Sébastien, who had big plans. Sir Jackdaw hadn't shown up. There were also a few contestants from southwestern France, accents in tow. The award ceremony, with English and Spanish sprinkled with Béarn accents, made this one of the rare occasions when Picardy mixed with the rest of the world.

I don't remember what I imitated that year—certainly not a bay bird, given that the evening lessons I'd requested had been turned down. I do remember, however, that Johnny chose to do the blackbird. He had a whistling technique with his fingers that produced an absolute, unadulterated sound. When he performed, the audience couldn't get over it. His blackbird melody was, in my opinion, a complete flop, but the jury and the audience didn't seem to share my opinion, and Johnny, who looked so small out there onstage, was the easy favorite.

I asked him how he whistled with his fingers, but he didn't want to explain. Sébastien, who used the same method, was impressive, a real master of waders. The guys from the southwest, with their imitations of goshawks, came in last. Up here in the north, you better have some wood pigeon in you!

Sébastien came first, and Philippe, from Mont-Saint-Michel, who also whistled with his fingers, came in second. I finished third, though I was first without fingers! And what about Johnny?

The emcee took the mic: "The winner of the under-sixteen category is... Johnny Rasse!" Johnny joined him onstage.

"Johnny," Pierre Bonte went on, "is that your real name?"

"Yes," Johnny replied, with a small, embarrassed smile.

"Okay then, Johnny-for-real! Your parents must have been big Johnny Hallyday fans. If you'd been a girl, your name would have been Sylvie, I guess!" Sylvie Vartan was, of course, the other half of the fabled French pop power couple.

Bonte laughed at his own joke, and so did the audience, but Johnny didn't crack a smile.

I didn't get it. The jury had given him the prize for the best bird caller under sixteen. I had come third overall, and I was twelve years old. Shouldn't I have been the top under sixteen? No, I was told; new rule. You couldn't place in more than one category. I was outraged.

I shuffled offstage with a heavy heart, thinking about the emcee's jokes about Johnny's name. Johnny was my nickname at home. The same name as my best enemy. That was what my parents had been calling me since I was little. They must have been Johnny Hallyday fans too.

If I'd been a girl, maybe they would have called me Sylvie.

Geese!

MY STRANGE TALENT BLOSSOMED during the year following my first competition. I started attending the Baie-de-Somme middle school, where most of the students were the children of hunters. Docquincourt, Delabie, Servant, Lamidel, Houart, Boyard… All their fathers were carpenters, fishers, or laborers, but every name called up a well-known local figure. The place had a reputation, too: it was a rough school, and the students were known to be troublemakers.

It was the first Friday in September. It was freezing, like winter had come far too soon. I got off the bus from Arrest and walked into the huge schoolyard. I lined up with my classmates, waiting for the French teacher. We could see our breath in the air, and puffed it out on purpose, pleased with ourselves. It looked like we were smoking behind the school, but without cigarettes. All the sixth graders were picturing themselves with cigarettes dangling out of our mouths right in front of the teachers. We laughed, forgetting for a second how cold our feet were.

The teacher came to get us, and we followed her to a prefab building with tall windows. In the classroom, we stood behind our desks as usual, waiting for the teacher's signal to sit down. Suddenly, one kid, Docquincourt, who was as short as me, broke the silence: "*Ozhons*," he cried in the local dialect; geese! He rushed to the windows and threw them wide open despite the cold. Some of the students didn't get it, but a few joined in, blocking the bewildered French teacher, who was begging them to close the windows: "What's going on here? Please, close them! I've never seen anything like this!" She threatened detention, warnings, and the principal's office, but to no avail. "For God's sake," she exclaimed, "what are these 'zonzons' you're going on about?"

"Madame," came the reply, "they're *ozhons*! Look at the geese!"

Through the windows we could see a large V of birds. A few of the boys started making a strange sound, and another. Soon, half the class was whistling, fingers in our mouths, hoping to draw and divert the geese. From the building across from ours, the ninth graders joined in.

Gradually, the students in the other classes started opening their windows too, intrigued, and soon the whole school was a total cacophony, our insane chorus echoing around the empty schoolyard and up to the huge flocks of wild geese circling hundreds of feet above us. Human geese called out from every window in the school. Some were whistling, others honked… Then the miracle happened: the flock of wild geese turned and answered.

In the middle of that French classroom that day, I felt like the school was welcoming me in its way. Graylag geese will respond to a kind of droning whistle, and the wild flock was shifting its trajectory in response to our entreaties. When humans vibrate our vocal cords and whistle at the same time, the noise we produce doesn't quite sound like a goose's, but it opens up an exchange that allows us to communicate with the birds. That is the secret of the graylag: it will only answer to that strange call, a combination of whistling and singing.

So it was that in the morning fog about fifty wild graylag geese dropped altitude and headed toward us. Seen from above, the blue-black asphalt must have looked like a giant pond. The geese had obviously mistaken our schoolyard for a vast body of water. When they got about thirty feet above us, legs out and ready to land behind the concrete ping-pong tables next to the cafeteria building, they finally realized their mistake. With a flap of their wings, they turned around and quietly headed off again, shrieking a fading goodbye.

We were all spellbound, teachers and students alike, amazed by the birds, their beauty, their wingspan, the majestic flight. Watching the wild geese go by always left us feeling a little nostalgic. Maybe they reminded us of all the beautiful things in life that delight us and then disappear forever in a flutter of wings.

The next morning, the classrooms were almost empty. In this unusual school, during the great migrations, children went hunting at night with their fathers.

The whole school year held the magic of that flock of geese. The school was wedged between the tall buildings of the old town and the vastness of the Baie de Somme, and as the seasons passed and gave way to other migrations, for four years, I felt the breath of the bay here, filling my heart and my lungs, my childhood.

When winter came in earnest, during the Christmas holidays, I was back to spending my days traipsing over the white pastures around Arrest. The Avalasse had slowed to a trickle weeks ago already. I played at sliding on the ice, carefully scuffling my way along. It felt safe: we'd had temperatures below freezing for over ten days, and the river was solid. A little ball of bright red feathers appeared, as in the middle of a white painting—a bullfinch come to say hello. The song of the male is so mournful that it always reminds me of those brothers from Le Crotoy who fell asleep whistling at the competition one day. It's such pristine sound, and so sweet, like a lullaby hummed softly into a child's ear. I struck up a conversation, and the bird answered. It was quite curious, and even accompanied me for a while along the Avalasse, flitting from tree to tree. I roamed across pastures undetected, walking along the center of the frozen stream. The steep riverbanks protected me from the wind.

Suddenly, I heard a strange cry—a human cry. It was far away, but it seemed to be coming from my favorite grove. There was no doubt about it, someone was calling me. The sound changed slightly—still far away, but I felt like it was meant for me. I climbed away from the frozen stream and scanned the pristine white expanse. There was no one there. The grove was about a hundred yards away. The cries sounded again: powerful, eerily human groans. The intonation intrigued me. Was someone asking for help? Calling to see if there was anyone there? The sound was obviously coming from the grove or, actually, upon reflection, from behind it.

Haaaa! Haaan! The noise made my blood run cold… I couldn't quite put my finger on it, but it was deeply distressing. I was heading to the grove when a huge graylag goose flew out at me, wings outstretched. The scene was worthy of the famous plane sequence in *North by Northwest*. The goose must have been aiming for the frozen Avalasse as its landing strip. It passed so close to my head I could feel the rush of air in my hair, and I could see its belly feathers moving in the wind. The image was striking. Its bright pink legs stood out starkly against the beige of its black-spotted belly.

I huddled at the bottom of my makeshift trench, my heart racing. The goose lifted back up into the sky, having decided that the Avalasse wasn't the spot it wanted after all. I thought about what I'd read in those bird books tucked away in my parents' bedroom. Geese are very sociable and live in family groups. When it starts to get cold up in Scandinavia, and the ground hardens and food gets scarce, ganders will urge their flocks to leave, leading everyone on a thousand-mile journey south. Up in the air, flocks are highly organized. Young geese line up single file in a V formation, with the gander at the head, choosing the best migration routes. The V system is not infallible, however: when it's foggy and they're tired, goslings can get lost, sending the group on a wild goose chase.

I recalled the episode with my sixth-grade class—that strange whistling sound that made the vocal cords vibrate, each student taking a deep breath before singing. The inhale wasn't in the chest; it was down in the belly. Without thinking, I tried the same thing now, placing my fingers in my mouth as if to whistle. I drew air into my lower lungs, swelling my belly, thinking of nothing else, and shouted, *Haaaa, haaan!* The cry was a primal, animal sound. It came from my gut, and made my head shake. Both the vibration and the breath were so powerful that the whistling was triggered almost simultaneously, creating a two-tone sound reminiscent of Tibetan chants.

The goose was already high up, its huge wingspan looming hundreds of feet over the ground. It dove. A gust of air behind me ruffled my hair. I lowered my head reflexively, then looked up at the sky. All I could see was a huge black shape and outstretched wings, nothing more. I tried again to make the sound that had made my body tremble. The goose responded immediately—it was like a lash. It seemed so close. I didn't move. I stopped breathing, stunned, and leaned against the riverbank. I could hear the goose scuffling above me in the snow. The cold grass crunched under its webbed feet. I crouched even lower, pressing my body against the frozen earth.

Still holding my breath, I looked up without moving my body or my head. I caught a glimpse of the goose's breath in the air, like a wisp of smoke that vanished immediately in the cold. There it was. It waddled forward a little more, and the orangey pink of its half-open beak let out another cloud of warmth. I was close enough to make out the little teeth that allowed it to munch on the grass it loved to eat, an indicator of its typical diet: graylag geese feed on fresh grass around ponds. I was less than three feet away.

I could feel that it was haggard, and lost. It shouldn't have been alone. Its head was bent, as if it was waiting for something or someone. Suddenly, it took off, landing again about sixty feet away. I inched over quietly for a closer look. It was a beauty, and I wondered for a moment whether there wasn't some sly pleasure there as it showed itself off in the field, knowing I was watching. I tried to make a slight sound, as if to try to tame the goose. My muted song, carried by the wind, grazed its body. It straightened up and, with its beak half-open, cried out once more, the sound of hope renewed. I was sure the goose was lost. I whistled a simple droning song. Its response was immediate, and intense. The tone of the song was unmistakable: the goose was calling out for help in a strange world. A few snowflakes fell, a subtle gray meteor shower. Our voices intertwined, the goose and I, piercing the white sky. Then, all at once, with a huge racket, ten, twenty, fifty geese swooped down to join the poor stray.

I could still make out my goose among the others. It stood up and honked loudly, happy to be back among friends. It sang a reunion song, a bit like deer bleating at each other in the woods. The other geese were making little contact calls. They recognized each other, saying hello as they took a cheery break in the Arrest countryside. The rest stop the stray goose had happened upon seemed to satisfy everyone—a happy mistake that led to the discovery of a new spot along their migration route. Finally, the gander signaled hoarsely that it was time to get going, and they were off, southbound again, for thousands of miles.

If only I could have jumped on the back of my goose to fly across the world up there, and get to know them a little better. I was so sad that day in December, and at once elated: sad to be so quickly parted from my goose, and glad to have heard the song of a beautiful, lost bird, a secret code that would become my way in among the children of the Baie de Somme hunters at school.

Pink Hawthorns

THE BIRD FESTIVAL WAS HELD in late April each year and concluded with the bird-call competition. Those days, Jean hardly ever came by the house anymore. A feeling of mistrust had crept in between us. He clopped around in his rubber boots, staring at me from afar. I watched him bike over to Jean-Pierre's farm. The farmer, a man in his fifties with huge hands and a face burnished by the elements, was Jean's best friend. Every day, Jean went out to his place. Sometimes he climbed up on the tractor, a brand-new John Deere.

Around the same time, I unwittingly befriended Jean-Pierre's brother, who lived three houses down. Gilbert had a face as wrinkled as the bark on an old oak tree. He was a cabinetmaking poet, and a beekeeper, a consummate nature lover. He knew how to graft trees, he made little mills that he placed along the Avalasse River, and he built wooden birdhouses, out of poplar for the people he loved, and fir for everyone else.

He had a workshop in the courtyard of Jean-Pierre's farm, with beautiful antique tools and enormous skylights that flooded the space with light. Every time I walked by and saw Gilbert working I stopped in, just for the smell of the wood. I loved his organized chaos, the tools hanging everywhere and pots of varnish on the table, the battered workbench. Now and again, Gilbert gave me a nesting box… made out of poplar, of course. Best of all, he was a wise old man, with so many secrets to share, and some of the keys to unlock their mysteries.

One day, he was explaining to me why you shouldn't use too wide a brace to drill the hole for a nesting box. Different birds require different diameters; the type of tree is important too. Blue tits like a one-inch, perfectly round hole. Redstarts prefer a more spacious nesting box, with a larger, oval entrance. How did Gilbert know all these things? His answer was always the same: nature, observation, and patience. Gilbert was something of a hermit. He went from his house to his workshop, from his beehives to his garden, and from his orchard to the Avalasse. His paradise was just a few yards from his home. Every season, every moment of the year was an excuse for some new tidbit of knowledge or new anecdote. Walnuts were harvested to make wine, apples for cider, and pumpkins for soup. Honey, meanwhile, was the elixir of nature's divinity.

Gilbert knew I loved birds, and that I was a bird caller, but he was thoughtful enough never to ask to hear me. While he was working, once, he stopped and said to me, "I like you, you're a good person." Then he picked up his plane and brushed away the cherrywood curls. "Jean I don't like," I thought I heard him mumble. "He's a know-it-all." I didn't understand where his sudden opposition was coming from, but at that moment it was clear that he was fully aware of our talents.

On the day we were leaving for the competition, Gilbert gave me the most beautiful gift. As we drove on the road where the two brothers lived, along the Avalasse, a shower of pink petals tumbled onto our windshield, just like in a Japanese film. For almost a mile, the road to Abbeville was draped in pink. My mother's eyes gleamed with happiness. My father didn't quite get it, but he saw the flowers as a good omen. To wish me luck, Gilbert had grafted all the hawthorns along the river, causing this avalanche of pink blooms to rain down on passersby.

That evening, Monique and Jean-Pierre had come to support Jean, but Gilbert stayed behind to boil his spinach and the first stalks of rhubarb. Nature waits for no man, and the harvest doesn't either.

The competition was held at the Hôtel de France. I was the youngest again this time. There were over thirty contestants that year, and the organizers had imposed a mandatory bird, the wigeon, always a popular favorite.

The rest was a mystery. Which birds would we imitate? Our secret plans were always a matter of some tension between Jean and me. A couple of weeks earlier, he'd told me he would be imitating the snow bunting and the common merganser, two species that were totally unknown up here. As for me, I took my father's advice and said nothing at all, telling Jean I still hadn't made up my mind.

Boots

JEAN OR JOHNNY?

Each one of us had our fans. In the months leading up to the competition, tensions ran high. To be honest, I felt like the competition day was my real birthday, my rebirth. In January, I would start pestering the festival organizers: I wanted to be the first to know the name of the mandatory species. And as soon as the results were in, I was on the phone every day, wanting to know when I'd be getting my check. I wanted to buy a new scope.

In between, there was the festival week. The opening ceremonies were held on the first Saturday evening of spring break, and the bird-call competition on the second Saturday. Every spring, there between the Rue de Catigny and the Rue des Abeilles, we played a fool's game.

I tried to spy on Johnny to figure out what bird he was practicing for the next competition. I could see his house from my bedroom window, but the road was so noisy I couldn't hear what he was up to.

Birds practice constantly too. Rehearsals begin as early as fall. The behavior has been noted especially for nightingales in their wintering sites: males will repeat their territorial song softly. In northern Europe, the following spring, they use the same song to seduce females and defend their territory against rival bachelors. On rainy days, you can often hear the muffled song of a male blackbird, tucked away in a clump of ivy. Then the weather turns and the same bird comes out onto the highest branch to belt his song confidently at the top of his lungs.

I rode my bike surreptitiously past Johnny's house, but I couldn't hear anything. Finally, one day, I found out that he was practicing the song of the common pochard. A bunch of local boys were playing soccer, and Johnny was with them, wearing a jersey and cleats from a trendy brand. I heard the kids calling me. "Boots! Come join us, we're missing a player!"

My nickname down in the village was "Boots," because I always had rubber boots on. The second I got home from school, I slipped on my boots, which were like old mail-boy boots, what they call seven-league boots in fairy tales. As soon as they were on my feet, the world was my oyster! It was such a pleasure to feel my feet slip inside, the cold rubber wrapping around my calves. I was a gangly kid, and by the time I was thirteen, I was already wearing size-fourteen shoes, so that my gait and the slap of my boots on the asphalt announced my arrival wherever I went in the neighborhood. But once I was traipsing through the swamp, in the mud, my boots were silent, and I could look up from the cow patties on the roadside to take in only the sky and the clouds: the freedom of not having to watch where you're walking

One group of birds are particularly well suited to marshy and muddy areas: the shorebirds in the order

Charadriiformes. The boots they wear are their long, thin, spindly legs, which allow the bird to stand as high as possible. Whenever water is deeper than their legs are long, the waves brush the lower abdominal feathers and right away the bird rushes to higher ground, onto the shoals, like a child who suddenly can't touch bottom.

Their legs are actually more like stilts than boots. Waders' legs can range from only a couple of inches long, for the dunlin or the common sandpiper, to about eight inches, for the black-winged stilt. Having legs twice as long as its body gives the stilt its typical elongated silhouette in flight. With its neck stretched out, it can measure up to three feet, yet it weighs only six ounces.

But to my mind the most astonishing shorebird is the stilt's little cousin, the elegant avocet, with its black-and-white plumage and its fine, upwardly curved beak. Its legs are about five inches long, and, unlike other waders, the avocet has webbed feet, which means that if the water gets too deep, it can continue its journey, walking along the shore or swimming for a short distance. Its feet are only semipalmated, so it has to kick really hard to swim at all efficiently, chirping its pretty *klooeep klooeep* or a fluttery little growl as it does.

With boots, anything can be a menace—a misjudged ditch, a muskrat den, or an awkward leap over a creek and the boot gets stuck and silted in and fills with water in a matter of seconds. Quick, you haul at the rubber on either side and try to suck the trapped foot back out, spreading your toes as wide as possible inside the boot to prevent it from slipping off and disappearing into the muck forever. A few good yanks and the boot gradually emerges, coated in black, foul-smelling ooze. At best, you'll get away with just wet socks; at worst, your boots are full of water and the way back home will be pitiful indeed, to a squishy *flock-flock* soundtrack at every step. That smell is the smell of defeat, not to mention the two or three days of moping around until the boots dry out and you can get back out into the marsh. Later I'd learn the trick of stuffing newspaper into the bottom of the boot, which cuts drying time—and being miserably housebound—by at least a day.

I love rubber boots. I used to wear them from winter to early summer. They were my signature. They were my passport, my way out into the world.

So there I was, playing goalie in my boots.

Johnny was in fine form, kicking them in one after the other. On his seventh or eighth goal, I asked him what he was working on, and he told me: the pochard. I rejoiced silently: the female has such a ghastly, hoarse, gull-like growl that it sounds like she's choking, and the male has a really nasal whistle. What a lousy choice! The Baie de Somme guys, with their almost heraldic wigeons, were going to eat him alive.

As for me, I was out fishing on the Somme canal when I discovered the bluethroat, *Luscinia svecica*, a tiny bird that was just beginning to settle in the Picardy coast at the time. The bluethroat is a marvel. It looks like a small robin, but with a blue throat. Its speculum feathers, in the heart of the blue patch, are a small touch of white that can only be seen when the male sings. The bluethroat's tail is dark and prettily edged with red patches, and it tends to hold its tail in the direction opposite its upturned head. You can spot these passerines in the distance, perched on willows or hawthorns, shaped like little two-handled vases, their tails stretched out and breasts puffed out.

Bluethroats have to be approached very gently. Most of the time, the bird will startle and dive straight into the nearest reeds, hiding until dusk. But sometimes it seems to tolerate the presence of humans, and you might even see a courtship display: the male soars into the sky, tail out, chirping, then he stops flapping his wings and parachutes back down to within a few inches of the female, who's completely camouflaged in the reeds.

I like the bluethroat's complex song, which sounds not unlike the great tit. That series of short trilled high notes is hard to get right, but it worked with my whistling technique. The next bird festival competition would pit the bluethroat's subtle high range against the growling pochard.

On the day of the competition, however, when Johnny's turn came, for the first time, he imitated the nightingale, one of Europe's best singers. He'd fooled me! My bluethroat was in the same family, the Muscicapidae, though obviously a less eloquent singer. Johnny's technique was peerless, his index and ring fingers supporting his tongue at the height of the exhale. He used his hands both as a sounding board and to modulate his airflow. The sound was clean, and he even nailed the incredible final decrescendo. But I felt like he didn't respect the bird's song. He never gave us the silvery, slightly rising melody. He didn't keep the unchanging order of the long notes in the territorial song of male nightingales, nor the sequence that follows, which consists of a few trills, and endlessly varied ornamentation. The ornithologists on the jury appreciated my bluethroat, even though I didn't do the best job. But as far as the audience was concerned, Johnny's nightingale was the best singer—his winning bird, which would take him places. Fortunately for me, my herring gull saved the day, and I ended up in the winning trio once again.

After I got my check, we asked around to find out what scopes ornithologists liked, and then my father took me to Belgium to buy my spotting scope. We converted the six thousand francs, my winnings from three competitions, into Belgian francs. (The postmaster in Arrest who made the exchange must have wondered what on earth we were up to.) And I finally got my Optolyth TS 65, with thirty times magnification. It was a gem! As an advance on my next Christmas present, I also got my first tripod, though it was so unstable in the wind that it fell apart the following winter in the Baie de Somme. As soon as we left the store in Bruges, we headed straight for the port of Zeebrugge to go see the terns and plovers that nest there, amid the shipping containers and hulking bulldozers.

Looking at the world through a lens allows you to see it anew. Along with the details of what you see, you feel like your senses are magnified. Seeing an animal up close like that makes you feel like you're suddenly at one with it, and yet you're hardly disturbing it at all. The bird is calm, it doesn't feel like it's being watched.

I saw the most beautiful things with my new gear. I felt like I could see as well as birds could, like maybe I even had birds of prey's double fovea—a small zone at the back of the retina with a high concentration of photoreceptors, where vision is most precise. Humans have one fovea in each eye, but birds of prey have two, giving them the best visual acuity of all vertebrates; they also have the added advantage of panoramic vision, and the ability to follow two prey at the same time. When I look through the scope with my right eye, and my naked left eye looks straight at the sky, I can see both the bird in its habitat and the same individual nearby—a kind of augmented reality.

I've gone through many spotting scopes since, but my first one still sits above my bedside table. Maybe it's waiting for someone.

The Nightingale

THE ABBEVILLE BIRD FESTIVAL had become the season's hottest local event. Each year, our spring break coincided with a reunion with my rival. This time, the draw landed me at the very end, while Jean was going second. Let the duel begin! At dinner, I sat near the door; Jean sat opposite. I took the elevator, he went for the stairs. As the contestants paraded through the streets, he was at the head of the procession, while I trailed behind.

The evening began with Denis Cheissoux onstage, emceeing. The first contestant performed, then it was Jean's turn to do that year's mandatory species—a challenge for him, since they'd picked the wigeon. The male has a loud, distinct whistle, which wasn't Jean's best register, though his voice was versatile, and he finished with a flourish, imitating the female's characteristic whistle. Many contestants would have a hard time with the wigeon: the male's call requires a dense whistle, almost like a singer's voice, with random variations and a strong treble. The wigeon showed me off at my best, with extreme vocalizations, a powerful breath of air, and a high C worthy of a coloratura soprano. It's a somersault off the edge of a cliff, requiring release, breath, and solid whistling technique.

My turn came, and I stepped forward, focused, facing the thousands of eyes staring at me, like a samurai ready to brandish his katana. I visualized splitting the air around me in two in a single gesture and with a single breath. I filled my lungs to channel my energy, and breathed out gently to center myself. I could feel the air drop into my lower abdomen. In a fraction of a second, my belly emptied, lightning fast, and a crystalline whistle sliced through the theater. Then—not a sound. Total silence. And suddenly, thunderous applause. Which was against the rules, of course, and Denis Cheissoux called for calm, but the audience kept clapping. My Baie de Somme whistler had set the room on fire. People stayed standing for a long time.

For the open competition, this year, the organizers had added the novel feature of projecting a photo of the bird chosen by each contestant. When Jean came up to the mic, instead of the elusive snow bunting I was expecting, the huge white mass of a herring gull appeared. I was sure he'd be imitating the little northern passerine, but instead here came his gull again, with the crowd, delighted as always, cheering him on. But this time, his act seemed different—less daring somehow, as if the gull couldn't quite take off. His voice was still a child's, but I could hear a veil in his throat, like a tightrope walker was wounded and wobbling. At the end, Jean bowed strangely to the audience, even though that was forbidden too. His performance had thrown him. The audience applauded warmly, but something was off.

It was my turn again. On the screen behind me the photo of a bird appeared whose name none of the other callers even dared to utter. I hadn't had the courage to tell Jean, despite his insistence; I was afraid he'd laugh at me. The bird I'd chosen lives in hiding, the Cyrano de Bergerac of the marsh. It is slightly larger than a robin, with dull brown feathers, and each spring the male cries out in despair, declaiming verses that its beloved can't hear. If he succeeds in mating, he stops singing until the following spring. A cursed poet of the undergrowth, it lives on almost nothing, but its song is legendary. When Denis Cheissoux named the species—"nightingale"—a shudder ran through the crowd. I could feel the vibration through my feet. The wave subsided. I hadn't yet made a single peep. I stretched my arms along my body. My hands delicately touched my lips. I thought of Gilbert's pink hawthorns along the banks of the Avalasse. I walked there, along the stream, and every year in that hedge I heard the nightingale's distraught song. Deep in the thicket, in the first warmth of springtime, its beak open wide, it cries its first notes. With each new strain, the nightingale tunes its instrument, beginning with a suspended gush of four or five long notes. I placed my fingers in my mouth, stood in front of the microphone, and exhaled a first wisp of air, so pure that at any moment it could break the musical line I was after and make me falter. This was a risky exercise. The resulting note was sculpted, miraculous. The balance between the air expelled, the position of the fingers, the placement of the tongue, and the inclination of the hand was tricky; that was what would determine the pitch, and the well-known territorial song. I did it five times in a row.

The nightingale's territorial song is a message to the whole swamp: *I am the nightingale, this tree is mine, and I humbly ask any female who lands here to be mine too*—a declaration so melancholy and tender that, once those five notes are out, this poet of the shadows can sing his true song, reveal himself. Like the rings of a tree, the number of trills indicates the age of the bird, and its sexual maturity. Long journeys between Africa and Europe have allowed it to develop its air sacs, syrinx, and feathers, assets that can help older males win their fair lady, outsinging younger opponents.

At the microphone, I embarked on the nightingale's great journey, pushing the air out of my own lungs, which had become air sacs for the night. The first trills rang out in absolute silence. A second salvo. The song was longer now, more intense. I imagined I had flown across the north, passing through the Strait of Gibraltar and the dunes of the Sahara before arriving at daybreak in Garamba National Park, in Congo. My first migration.

For the third verse, I performed a few trills, heralding a seasoned migrator returning one last time to the hawthorns in Arrest. My eyes were filled with the sand of the dunes, the salt of the Mediterranean, and the drizzle of the northern sky. Slowly, I moved out of the light. The audience was silent, contemplative. The love song had gone beyond human language, opening all of our eyes to the sacred beauty of nature. For me, it was a turning point, the beginning of a new era. I felt consideration; I felt deep gratitude. The call of the wigeon and the song of the nightingale showed my range, and my ambivalence, caught between the old traditions of the Baie de Somme and the artistry of the pasture's fine singers.

This third bird, for the last round, wasn't going to change a thing. The competition was over. I went back to my seat. I could see Jean across from me, slouching in his chair. He wasn't moving. I didn't feel like going out to see my family or the audience like I usually did. I was thinking of my birds. The previous year's winner, who was this year's judge, came to say hello. I would be the winner tonight, he told me.

It was going to be an unforgettable evening, but Jean's despair and perhaps the melancholy of the nightingale were getting to me.

Standing next to Denis Cheissoux onstage was a young blond woman draped in a Miss Tournai sash and holding the first-prize trophy depicting two gannets. Before awarding the prize, Cheissoux announced, there would be a surprise guest: "Give him a round of applause, ladies and gentlemen… Zorro!"

I was clinging to my chair, adrenaline surging through me. Was Zorro there for our duel? If there was a duel, I didn't stand a chance. I imagined a masked man in black boots and a black hat impaling me on his sword.

Zorro strode onstage with a swagger and a big smile. He looked like the actor Alain Delon: tall, blue-eyed, forty-something. He wasn't wearing anything remotely swashbuckling; rather, he was dressed up in a suit, as if he'd been waiting for this moment for a long time. He was limping and strutting at the same time, which had a weird effect. He circled his arms to rile up the crowd, and demanded the mic. The audience laughed, and Cheissoux obliged. From that moment on, Zorro was the emcee. In an accent so thick you could have cut it with a knife, he rattled off jokes and stories about birds and bird behavior. He was the undisputed champ of anthropomorphism. He roasted local politicians. He was recounting his love of hunting when he suddenly noticed the presence of poor Miss Tournai. The young woman immediately became the focus of his routine, though she had clearly done nothing to indicate interest in participating in any way.

Zorro improvised a little courtship ritual. All the Columbidae were represented: Eurasian collared dove, turtledove, African collared dove, wood pigeon, rock dove. Denis Cheissoux tried in vain to retrieve his microphone and put a stop to the nonsense, but Zorro jumped aside, laughed, got away. His last attempt at seduction was the courtship dance of a male turkey. Head down, he gobbled and charged at the young woman. It was too much. Miss Tournai left the stage.

Then Zorro decided to do a few birds, in decreasing order of size. He started with geese, ducks, and waders. He tried his hand at song thrushes, blackbirds, a few chickadees. His technique wasn't perfect—I could hear some air in his whistle—but the series of species itself was a performance. I was intrigued by his fingerless technique, the way he used only his lips. From my seat, I could see the placement of his teeth and mouth. You had to be smiling to be able to whistle like that. While I used my fingers and hands as tools, almost like a syrinx, Zorro could whistle without hiding behind his hands. His method was great for speed and tonal change, and worked especially well for certain species. And Zorro remained the first hunter who dared imitate passerines and songbirds. Some called him "Bisou"—little kiss—and he was "Zorro" to others, a local legend. As a hunter, he always skirted the edge of legality. He had a way of defying authority, a kind of Baie de Somme Raboliot, like the famous poacher in the well-known novel.

The real emcee finally got his microphone back, and Zorro bowed off the stage as his fans cheered. Miss Tournai returned and the awards ceremony resumed. Second place in the all-around went to Jean Boucault. Jean smiled as best he could. Unsurprisingly, the nightingale had worked its charms, and I won first prize in the under-sixteen category. I could

My brother was in trouble again. This was the third time he'd come crying to me because someone was punching him in the face every time he walked into the schoolyard. To make matters worse, his bully was the new kid nobody knew but of whom everyone was wary: word was that his father was a former champion boxer turned fisherman.

He was already a solid six feet tall at the age of fifteen, and he would always wait at the same spot by the school gates, probably ready to escape or to run away. I had no choice. My honor as an older brother was at stake. I had to do something.

So I decided to fight the boxer's son from Le Tréport. I was ready to give my life there in the schoolyard to defend my baby brother.

At recess, without any preamble, I stepped forward and drew first.

"Hi! I hear you've been beating up my brother?"

His back was to me. He answered without even turning around.

"What, you want a piece of that too? Who asked you, anyway?"

I had seen a Clint Eastwood clip in a French TV show about classic American films, and I knew this didn't bode well. But somehow my mouth short-circuited my brain.

"I'm not going to tell you twice. Stop it!"

I didn't even see the punch coming; I just felt a whoosh of air go past my nose. He missed, but it happened so quickly I was surprised that I was still standing.

In seconds, a crowd formed around us, our classmates descending like vultures.

"Fight! Fight! Fight!"

Someone called out my name, cheering me on. "Come on, Johnny!"

The colossus stopped dead in his tracks. "Johnny? You're Johnny?" He grabbed me by the shoulder and threw his arms around me. I was dumbfounded. He was hugging me so tight I could hardly breathe. That's when I found out he was the nephew of a contestant from the previous year's bird festival.

"This is Johnny Rasse! It's Johnny Rasse," he was shouting now. "The nightingale, the blackbird! Don't you guys know? He's incredible!"

Who was my brother, he asked me. A procession was organized, and the whole school followed us as we set off, arm in arm, in search of my little brother. When we finally found him, he apologized profusely, and pledged to protect us. And he confided that if he'd landed that first punch, his father would have kicked his ass so hard... As far as his family was concerned, I was the best bird caller ever, even better than Zorro.

That was how Johan became my bodyguard in the school jungle. Until I graduated, every time I won a bird-call competition, my popularity swelled. It changed my life. The high point was in ninth grade, when the principal dragged me out in front of a microphone in the middle of the gym to perform to the other students. After three formal requests had been sent home, I was forced to comply. You couldn't say no to the principal at Baie-de-Somme school.

The Kestrel

ALONG WITH THE START of grade nine came a minor revolution in our family organization: I became a boarder at Amiens, in a tiny attic room above the school. It was deadly boring. I missed my chickens, my ducks—I missed nature, and especially birds. Before boarding school, I had never felt time passing at all, while now, all of a sudden, it seemed to be crawling by. I watched the second hand ticking interminably slowly in study hall. Our French teacher had assigned Hervé Bazin's *Vipère au poing*. We had a month to read it. I went through it not once, not twice, but four times; I had nothing else to do, and so from a quarter to five to ten thirty at night, I read and reread that book. I knew the novel by heart, I could recite it. Before goslings reach their sexual maturity at the age of three, they have no choice but to follow more experienced adults. I was like a stunted little goose holding back the migrating flock.

Then came time to write our in-class exams. I handed mine in, but I never saw it again. I got in trouble for cheating! They said I must have looked at the book during the exam, and gave me three hours' detention. I hated French…

Thank heavens for our science teacher, Monsieur Nosal. The first time I walked into his classroom, on the top floor, I caught a whiff of a smell I knew well: a mix of decaying plant matter and feathers. The students rushed in, settling on the white tiled benches. There was no teacher. We waited. A door slammed, and an eerie silence filled the room. Suddenly, behind me, I heard the sound of a bell, and I saw a kestrel zoom by, grazing my face with the tips of its wings. It was a female; the male has a gray tail. The bird flew at full speed toward the chalkboard, and I fully expected to hear the terrible sound of a bird smashing against a window, but it veered around and flew back over the classroom from the first to the last row, landing on the arm of a bespectacled little man in a white smock, wearing a falconer's glove.

Lesson one: the kestrel. I couldn't believe it. I started looking forward to Thursday mornings impatiently. Whatever I handed in in his class that year consistently came back marked "A. Admirable work!" I'm grateful that my classmates didn't hold my weekly two hours as teacher's pet against me. After class, I spent hours with Monsieur Nosal in his classroom, looking at slides of deer, woodpeckers, all kinds of snakes. Along with his wife and Laurent, an ornithologist friend, he organized outings on Wednesday afternoons, which we had off, and we went bird-watching around Amiens. He even signed me up for a bird-watching trip to the Danube delta. But first, I still had a whole winter at boarding school.

I used to finish all my homework on Thursday evening so that I didn't have to crack a single book from Friday to

Monday morning, and I could enjoy being home in Arrest. There was so much do in the village. At last, I was back on the Avalasse with my birds, and back on the farm, to the comfort of seasonal work.

The boarding school was mixed-level, from the last year of middle school to the end of high school. The younger students had less homework, and we often got bored in our rooms after supper. Some of us went to sleep early, while others, like me, spent long hours staring out the window. One evening, the night don, who liked me, let me leave my room to go listen to the birds. I had half an hour, and not a second more. I snuck downstairs in the dark, first down the narrow wooden steps from the attic where my room was, then the two wide black-and-white speckled staircases with the cold metal handrails. Strangely enough, inside the building I was scared, but my fear vanished as soon as I pushed open the door, which had to stay unlocked for safety reasons. In the schoolyard, the night opened up to me. I hid at the foot of a linden tree, waiting to hear the secrets of the pair of tawny owls that nested in the tall trees behind the school. I'd been hearing them for weeks, but my bedroom window faced the wrong way, overlooking Maryland, which may sound exotic, but Maryland was just the name of the slot-machine bar across the street.

I'd been waiting at the foot of the tree for a good twenty minutes, and I was kicking myself for not having dressed properly for the December night. My throat was starting to get scratchy. Just then, I saw a strange bird indeed: there was my Friday-afternoon teacher, quite out of her natural habitat, sneaking out of the abbot's apartment!

"Bonsoir, madame!"

The nocturnal encounter was providential, as it turned

out. The following Friday, I casually asked to leave my last class a few minutes early to catch the 5:04 train home. As I anticipated, my request was granted, and I didn't have to wait for the 6:12. The last hour of class seemed to last forever. As discreetly as I could, I kept glancing out the window at the bell-tower clock: 4:53… 4:54… Finally, I was free!

Lucky for me, the Rue du Collège ran downhill the half mile to the station, due west. Exactly where I wanted to go. Toward the sea. I ran into town with my backpack, as fast as I could, jostling a few passersby who had the lousy idea of not getting out of my way, but there was no time to explain. I locked my eyes on the Perret tower, which loomed like a beacon above the train station. I soared over crosswalks, flew down staircases, and leaped onto the platform. By the time I got on the train, out of breath and drenched in sweat, I didn't even have the strength to fight for a seat next to the hordes of other students on their way home. I huddled between two compartments on the old Corail train, rested my head on my bag of dirty laundry, and, on my weekly migratory run, I fell asleep.

A few days before undertaking their great migrations, birds go into a frenzy. In German, ornithologists call the behavior *Zugunruhe*, literally, "migration anxiety." Birds stuff themselves before the trip: a reed warbler, which normally clocks in at about four ounces, can double its weight in a few days. Their sense of direction naturally begins to take over, telling them where to go, and the obsession to migrate becomes stronger than anything else. Early experiments, at the beginning of the twentieth century, studied birds in captivity that had had their feet painted, which allowed scientists to trace movement patterns in the bottom of the cages. In every case, by the end of summer, passerines—wheatears, robins, or warblers—would begin to pace around the southern end of the cage and almost never venture into the northern part at all. Instinct guides their steps.

And then they're off. The birds seem unstoppable—even a sandstorm on the way in the Sahara, which kills thousands of swallows, doesn't make them turn back. Whatever the cost, they try to make their way across.

Song of the Cranes

I DIDN'T MAKE MUCH of an impression as a bird caller at boarding school. I hadn't said much to my classmates about my passion, and I knew I'd never be able to practice during the week. That meant I had only weekends. To make matters worse, the mandatory species that year was the common crane. The imitation had to be done as a duet with another contestant. It takes a lot of hard work to reproduce a crane's song: you have to breathe in deeply to get a clean sound. It's like a trumpet blast, but there's a juxtaposed throaty trill that modulates the single note. The trills are more or less pronounced depending on the kind of call you're doing. When it's well executed, the result can be magnificent.

The trouble was that cranes' migratory route was nowhere near the Baie de Somme; it was two hundred miles east, running from Metz to Biarritz. Common cranes on their way to Spain from Scandinavia gather around the Lac du Der-Chantecoq, a lake that evokes the rooster's song, though around that time of year the song of cranes is all you hear!

During our February break, I was getting ready for that year's bird festival. I had a recording of common crane calls by Jean Roché on a four-volume compilation of European birds. It was on the second CD, along with the western capercaillie, disc two, track twenty-three, but the recording was only a minute and twenty seconds long. I listened to it so often that I ended up scratching the disc. I simply had to find a way to hear cranes in the wild. Finally, at my mother's insistence, my grandfather said he would take me for the day, as a late Christmas present. That was my Europe 1 grandfather, who had divorced his wife over a radio station. Wherever he was, he always had a little transistor radio with him. He was always up to date with current events, before anyone else. He knew all about politics; he knew a lot of things, but he never had time to talk about anything, because he had to listen to the news…

After listening to four and a half hours of Europe 1, I got out of the car. The spectacle before me was astounding. There were cranes everywhere across the fields—larger adults, with contrasting gray plumage and hairless, reddish napes, and smaller yearlings in their dirty brown peasant outfits. In the afternoon, we changed locations. My book on where to see birds in France gave precise indications as to where to go. In the evening, the place to be was on the northwest shore of a lake—that was the magic spot at dawn, and at dusk, when the cranes, heading to bed from their daytime feeding grounds, would fly so low that when you were looking through binoculars it almost felt like you could reach up and touch them.

We were sitting comfortably behind the car, out of the wind. It was freezing. I had the scope out. Crane calls echoed everywhere in the distance. A northern diving duck was fishing on the small lake, a female common merganser. She wasn't bad-looking, with her milk-chocolate neck, white cheeks, and ash-colored body. It was the first time I'd seen one. I was happy, but I'd have rather seen a male: they were much prettier, if my book was anything to go by.

The calls of the cranes rang out over the countryside, the birds answering each other. Little by little, the sound seeped into me, an indelible note imprinting itself. I felt like a polyglot unraveling a language's hidden structure, like I could feel meaning soaking in. Slowly, the intensity increased, and the first cranes started flying in, back from the fields—first small groups, families. I could clearly make out the contact calls of the young, a kind of light croak, barely audible, the sound and strength varying depending on the individual. The leader usually had the most powerful song, conducting the choir, the others responding in kind. Tonight's show was going to be wonderful.

Just then, my grandfather called me back.

"Come on, Jean, we've got to head home. They're saying on the radio there's a risk of freezing fog to the east. The roads will be bad tonight." I reluctantly complied. As I nestled down into the comfortable seat of my grandfather's gray Renault 21, through the window I saw an enormous black line cutting through the sky like a knife. The great gathering of cranes was underway at the lake, though Europe 1 didn't mention that.

Things got interesting when school started again. One townie had bought cigarettes to sell to a senior, and, after dinner, we would meet up behind the gym to smoke. The night don didn't seem to mind the stench on us when we went back to our rooms. I managed not to cough, and I was feeling pretty proud of myself; good thing I'd been practicing with clematis branches, though the twigs didn't smell nearly as bad as the glowing red butt we passed around.

Our fun lasted about three weeks, until one night, from the other side of the gym, we heard someone shouting our names. Shit! It was Hitler! Hitler was the school monitor. He'd earned his nickname one day when he refused to open a window in study hall to air out the room after someone had lobbed a stink bomb in. Our night don was sick, and Hitler was filling in. The jig was up. We all got caught, and the next morning, the sentence was handed down in the guidance counselor's office: all the boarders were on the hook. They gave us all detention for the following weekend—the start of the April holidays! We would have to spend the Friday, Saturday, and Sunday nights at school, and our parents could pick us up on the Monday morning. We'd have lots of painting and repairs to do, Hitler promised. My parents were going to kill me. We had to bring home a form for them to sign, explaining the punishment and the crime. I could imitate a hundred birds; I guess I was about to learn how to imitate a signature!

Two days before our house arrest, as we were jogging around the linden-tree-lined avenue to try to qualify for the intercollegiate cross-country race, I saw the guidance counselor's car pull up. He drove a little white panel van, a Citroën C15—a real country car, which stood out in the teachers' parking at our big-city school. At the end of gym class, I ambled over to the car. There, on the rear window, was a sticker showing a gray partridge hiding in tall grasses. The guidance counselor was a hunter; he had to be familiar with the bird festival, which after all was organized in collaboration with the Baie de Somme hunters.

On Friday morning, after a long, restless night, I decided to go see the guidance counselor at ten o'clock recess. My parents still had no idea we were being punished; I absolutely had to go home that night. In the guidance counselor's office, I pleaded my case: "Do you know what the audience will say if I don't perform at the bird festival Saturday night? They'll say it was the school's fault... and the guidance counselor's!"

He was flustered. A week earlier, a *Courrier picard* journalist had come over to our house to interview me. They ran a full-page spread about how hard I was working to prepare for the event, two weeks before the competition. The article was photocopied and posted all over the school, even in the teachers' lounge, by Monsieur Nosal, who was chuffed to be my teacher.

"I've been practicing the gray partridge for the competition for over a year!" It was actually the common gray crane, but who was I to split hairs, or feathers. There I was in the guidance office imitating the call of the male gray partridge rounding up his hens home at sunset. The counselor's eye twinkled, but then he faltered. Suddenly he wasn't in his office anymore but loping through the beet fields alongside his Brittany spaniel. I scratched out a few more partridge calls, my best ever, I think. And then, the coup de grâce: "It would be like if you weren't allowed to attend opening day." He stared at his blotter as if it were a crystal ball. I'd hit it home! "All right, all right, you're free. But I'll thank you to make sure you win the competition!"

Aboard the crowded train bringing students from our region to spring-break destinations on the Channel coast, from Noyelles-sur-Mer to Calais, I watched the Picardy marshes whirr by through the window for a long time. The bird festival started the next day, but the competition wasn't until the following Saturday. I felt the butterflies starting up in my stomach.

This was my favorite time of year. The willows had bloomed, festooning the marshes with their sparkling yellow pompons, and now the hawthorns were daubed with new leaves and immaculate white flowers. As the trees bloomed and May drew near, a distinctive scent filled the paths, a mix of rotting fish and cat urine. One day, as I was walking between Saigneville and Boismont, I learned the name of that unique fragrance from an old lady walking her dog: "Can you smell that?" she asked. It was the smell of Jesus's diapers, she said.

I was no saint. My friends were all stuck doing manual labor at school while I'd been sprung free on false pretenses. I wore my crown of thorns all weekend. Saturday I was bored, and Sunday was no better. I wondered what my buddies were up to. The bird-call competition was only a few days away, but I wasn't in the mood. To top it off, those cigarettes I'd snuck, my first time smoking, were affecting my voice.

The program that year included the blackbird, the common crane, and the herring gull. I was sitting in the hall, waiting for the draw. A duet requirement had been added this year by the organizers, who were looking for something new. Johnny didn't look at me; he seemed to want to team up with a guy named Houard, from Ribeauville. Meanwhile, a twentysomething surf fisher from Le Crotoy, Cyril, was making eyes at me. I'd known him for a few years. He was a bay bird whistler who always ranked pretty low. I'd never seen him do any other birds, so I didn't think he could help me do the crane. But people already seemed to have paired off, by provenance: two guys from Le Crotoy, two from Saint-Valery, two from Cayeux-sur-Mer…

"Can you do a crane?" I asked Cyril.

The man from Saint-Valery was sitting next to Cyril at the table. "Not even a Poclain winch!" he cracked, invoking the French heavy machinery company, laughing uproariously at his own joke. But this was far too serious for me to even smile.

"Yeah, sort of..." Cyril replied.

"Can you do a bit of the song, and the trills?"

"Not really. But with you, we'll manage!"

"Okay, let's go for it. I'll tell you what we're going to do."

I showed him what I had up my sleeve: I'd gotten a can of red hair-spray dye through a townie from a costume store in Amiens. The color was a little too vermilion for the crane's neck, but with the lighting in the theater, it would do the trick.

Cyril looked at me, unsure.

"Yes! Listen, seriously. We're supposed to go fifth, out of seven. As soon as the duet round starts, we'll run to the bathrooms and spray our hair to look like the cranes' crest. It'll be a riot, you'll see."

So there we were, hiding in the bathrooms at the Hôtel de France. Two guests came in as I was explaining to Cyril about common cranes' courtship dance, from what I had seen a few weeks earlier at the lake with my grandfather and from the description in the *Audubon*.

"First we walk next to each other. I'll do the male, *grrrr-roooo*, singing, like I'm indifferent, and then you answer. Go ahead, answer me!"

"Are you sure?"

"Yes! Come on, answer!"

"*Krooweerooeekrrr...*"

"Actually, never mind, don't answer after all. I'll sing, and you keep walking by, and then suddenly, when I start in on the low notes, you turn toward me, lightly, like this, skipping, with your head down, and you come closer. When you're about three feet away, you straighten up. Okay? We'll stand facing each other and open our wings—I mean, our arms, and then we'll sing together at the top of our lungs. You sing too, but not too loud. Just do a little trill in your throat. But never louder than me. You're the female, okay?"

"Are you sure?"

After some arduous rehearsing, we were finally called to the stage. And… we won! Our crane dance was voted best duet. The staging had worked! I knew it would. I was overjoyed. Two thousand francs to share, a thousand apiece. A trophy, applause. We did an encore. Cyril turned out to be a very convincing female crane. The 850 people in the audience were clapping wildly. The houselights went up, and we took a bow. But I started: there, in the fifth row… Was that the guidance counselor? Sitting next to his wife? I barely had time to see anything before they dropped the houselights again. I thought it was him… There wasn't a second to lose. I had to get out front, grab my family, and get out of there first. I couldn't risk them talking to each other. I don't remember what excuse I made up, but that year we weren't going to attend the reception put on by the mayor of Abbeville, who always invited the competition winners over for some *gâteau battu*—literally a beaten cake, referring to how it is made—a kind of tall brioche, our local specialty.

In the car, on the way back to Arrest, I didn't say a word. Like the cake, I was beat.

Ballbusters

IN THE TRAIN BACK to school on the Monday morning after the break, I was nervous, but it seemed there was nothing to worry about: everything seemed the same as before. I asked the three boarders in my class how their long weekend at school had been, but they were evasive. The most I got out of them was a laconic "fine." I had so many questions: What had they done? Who'd supervised them? But they had nothing to say. At the evening meal, with all the other boarders, I tried my luck again: How was their Sunday at school? Again, nada. After supper, we headed back out through the gym for our usual last half hour in the schoolyard.

I tagged along with the group. Suddenly, everyone turned to look at me, and before I knew it, I was surrounded. My dorm mates ran at me. There were three, five, ten of them, coming in for the kill. I was swarmed, I was prey, but I had no wings. I struggled in vain. Caught and resigned like a sparrow in the talons of a hawk, I awaited my fate.

Most birds of prey tend to be solitary hunters, but in the Americas there's one species, the Harris's hawk, sometimes called a wolf hawk, that hunts in groups. A group of three to seven birds will hunt together, which allows them to be more effective, especially for larger prey. One group of hawks will flush out the animal, then others ambush it and take it down. Harris's hawks are appreciated by falconers, who use cooperative hunters for bird control, for instance. Like other European falcons, gyrfalcons in captivity will sooner or later kill their block mates, especially if they're smaller. And any pest-bird control protocol that tries to combine species tends to end up a giant brawl.

My friends carried me by my limbs, holding my legs and my arms spread wide, my head tilted up to the sky. I could feel us moving toward the goalposts, and started screaming as loud as I could. "Nooooo! Not that!"

The ordeal began. I don't know how long it lasted. Grabbing me like rowers, they rammed my crotch repeatedly into the metal post, over and over again. The pain was like an electric shock every time. Finally, when they tired of the back and forth, they held me and spun me around the cold, hard metal. I couldn't breathe.

Birds' gonads are internal. It can be difficult to discern the sex of immature birds, or of birds that lack secondary sexual characteristics, like colored feathers, crests, or calls. A famous case in point was the reintroduction of a breeding pair of bearded vultures into the mountains in France. No one could figure out why they weren't mating, until, five years after the release, ornithologists realized the birds were both males. In birds, only the left side of the female's genitalia develops, while the right side gradually atrophies. Both the male and female reproductive systems are seasonally sensitive: during breeding, their reproductive organs will swell considerably, up to ten to twenty times larger for females, and one hundred to three hundred times for males.

Out in the soccer field, I gradually came to. I was alone. I could barely make it up the stairs to my room. I walked right past the night don's door, and he ignored me. I could have screamed as long and as hard as I'd wanted, I don't think he would have done anything. I'd spent the whole weekend wanting to be with my friends, while they'd spent it planning to destroy me. And the next day, it was all over, and they were talking to me normally again. I was back in the gang. I never said a word about what had happened.

The Emperor and the Nightingale

ONE SATURDAY, A BIG LUNCH was organized at our place. Once again, my father had decided to show me off to his friends. He tried to get me to come out of my bedroom.

"Johnny... Johnny!"

That was how he started coaxing me off my perch.

"Johnny, come downstairs, Michel's here and he's got something to ask you!"

I was fully aware that Michel had nothing to ask me at all. He just wanted to see the strange creature who talked to birds.

But I stayed locked away in my cage. I refused to be anyone's pet canary, or a nightingale that would sing at the emperor's summons. This time, the war of wills lasted twenty minutes, five minutes longer than the time before. At last, I tiptoed downstairs, hoping that the friend had left and that

I could avoid playing parrot. Just a few more steps… I crept toward the front door, and freedom. Locked!

My father's ploy had worked; I was trapped. He narrowed his eyes and dragged me to the living room, pleading, a pathetic drama between a father and his recalcitrant teenage son. But I was forced to give in, and the act began as usual: waders, ducks, sparrows… Out of respect for that day's audience, I remained pleasant and polite. I had no choice but to play along.

The ritual was always the same. Every two weeks or so I was the main course at Chez Rasse, between roast pheasant and custard. My father acted as emcee, rattling off the names of birds. The house was turned into a grotesque cabaret for my father's friends, many of whom were hunters. Word of mouth did its thing, and more and more people came to see the bird-boy.

After the show, I couldn't bring myself to eat, and I slumped offstage, drained. For years there were few Saturdays when I could just enjoy my lunch, relaxed and carefree. I'd become a puppet, a windup bird my father cranked for his friends, and it squashed the happy child I had once been.

The year I graduated and moved on to high school was also my last year in the bird festival's under-sixteen category. I began using a technique inspired by Zorro. Whistling without fingers took me to unexpected new heights.

This time, I went wild, with the skylark, chaffinch, and blackbird. I sent in my entry and species selections before my father did. There wasn't a single one from the Baie de Somme, as if I were flipping the old pied piper the bird.

As for Jean, he had gone over to the other side. He was over sixteen, and a regular contestant now, a contender for the overall victory for the first time.

He'd changed physically: he'd always been gangly, but now he was even taller, and had grown a wispy moustache. Whenever I saw him, he would clear his throat. There was something wrong with him, or else he was hiding something from me, I could feel it. He avoided me as much as possible and never spoke to me. He seldom spoke to anyone at all. It was as if any contact with anyone else would betray the secret he seemed to be hiding.

There were eight of us in the youth competition, and twelve in the overall. The director told me that it was getting difficult to find contestants because of the bar Jean and I had set: newcomers were afraid of making fools of themselves. We couldn't let the bird festival become the Jean-and-Johnny show.

Denis Cheissoux was emceeing again and opened the proceedings. I drew the fifth spot. My technique was exemplary, and I seized the opportunity to do some high-pitched vocal acrobatics with the skylark. The constraints of my new technique, because I couldn't control the whistle with my fingers, also caused some slipups, though I felt like its potential was infinite, with great freedom and prodigious speed. I could aspire to small passerines now, and was already dreaming of blackcaps, chiffchaffs, and dunnocks. The audience was astounded by my new range.

Jean was going tenth. He was nervous: I could see the veins in his neck popping, and he kept swallowing. He looked sick. He came up to the mic.

He had decided to imitate a little-known but legendary bird, the holy grail of all reed-marsh birders. The bird had a name but was so stealthy and timid that he was almost never seen: the bluethroat. Jean's choice was clearly meant to curry favor with the naturalists who were at the festival that year,

Guy Jarry, Jean Dorst, and Philippe Carruette. He positively lit up whenever he had a chance to rub shoulders with the big names in ornithology.

The bluethroat's whistle was a little clumsy, insipid, muddled, without much meat, more like some kind of distressed swallow. Jean was probably picturing himself standing at the end of a reed, sticking his head out and showing off his blue breast. He caught the moment for a split second before it vanished and it was just him slouching up there onstage, all legs, collapsing into the bird's chirpy finale. The longer the imitation went on, the more striking the resemblance actually was: like Jean, the bluethroat was a fragile, ungainly thing with long, skinny legs, living a life concealed yet so proud of its blinding blue throat.

But tonight, the audience remained unmoved. No one knew the bluethroat. There was an awkward silence, and everyone seemed embarrassed by the precise but somehow shameful performance. Jean seemed disoriented and was backing away, looking down, when a voice cried out a resounding "Bravo!" shattering the stony silence, as if to say, *None of you get it!* It was Guy Jarry himself. In spite of the impassive audience, the expert's approval made Jean smile again—as if there, on the eve of his seventeenth birthday, with his rendition of a bluethroat bobbing on a reed, he'd been knighted.

Jean's performance asked us to consider what we really thought about bird calls and love—the fascination the songs of birds sparked in each of us, and the relationship between the audience and bird callers. Did callers really just have a duty to be pleasant? Could we imitate less-pretty songs? After all, disharmony and dissonance do exist, even in great musical works. Birds are no exception to the rule, and the hoarse cry of a gray heron or a griffon vulture stands in contrast with the eloquent beauty of a blackbird or nightingale.

That night in April, Jean reminded us of that tension. Personal taste, enjoyment, and the need to please must not lead us off the path of truth. The sincerity of a bird's song was also contained in its apparent dissonance.

For my second bird, I was counting on the cyclical song of the parkland master, zooming from a crescendo to a rapid, continuous decrescendo. The male chaffinch is the quintessential musician, with a universally recognizable, rapid, loud song. No one could imitate it—not Zorro, not Jean, no one. I was the first to try.

The song consists of two quick little twitches, like two electric shocks, one starting out high and the second ending low down, as if this little bird that weighed under an ounce had been holding air in its tiny, distended lungs for too long. The passage between these two twitches was tough, but I had it down. That was more than I could say for the end of the song, which I more or less made up every time. I couldn't for the life of me get those last notes.

In fact, the end of the chaffinch's melody was an ornithological enigma that has only recently been understood. Chaffinches have a territorial accent, which differs according to their place of birth, and which is their signature in the end of that phrase. There are huge deviations, for instance, between chaffinch songs in northern and southern France. I was banking on the fact that the hesitation in my song would reflect those regional disparities. The audience was so dazzled by the power and the nimble stunt-singing of my chaffinch that I was interrupted by a standing ovation.

Then it was Jean's turn again. Breaking from his usual practice, he was doing his herring gull second. He seemed agitated this time as he edged up to the mic. He stretched out his arms and let out his trademark cry. But the note the audience was hoping for shattered like glass. The crystalline cry they were expecting was speckled by tiny grains of sand scraping at the back of Jean's throat. He turned his head away. We looked at each other; his eyes filled with tears. He tried again. The cry was more sonorous, and deliberate, the graininess more diluted now in his incredible energy, but the hoarseness was still there. Jean was sick, his voice wasn't working, the sheen had come off his previous performances. Although the typical mewing contact calls were worthy placeholders, the clamor of herring gulls we all heard on the walk to school was no more than the fading memory of a vanished childhood. Jean decided to put an end to his suffering. He backed away, looking at the audience ruefully as if pleading with them to forgive him. The crowd was grateful that their customary gull had put in an appearance and jumped to their feet, unaware of the turmoil unfolding within.

Jean sat back down beside me, waiting for the last round. He tried to gesture something with his hands, but I didn't understand, so he leaned over to whisper in my ear. He almost couldn't speak at all.

"I can't do it anymore! It's over."

"No! The audience loves you!"

He cut me off: "You don't get it, my voice is changing. It's over!"

The world of our childhood, populated by gulls fluttering like kites in paradise between the Baie de Somme and the Avalasse River, vanished suddenly and forever in Jean's big blue eyes, still gleaming with tears. He'd fallen from quite a height. I wanted to tell him it was nothing, that it would be okay, it would come back. But already for a while he had been desperately clinging to his child's voice, fighting, practicing for hours on end, panicking at the idea of losing his birdness, his identity and everything he had ever been, trying to hold on to what was already gone.

It was time for the third round. Jean looked drained, his head bowed, gazing at a spot on the stage floor, as if he were trying to escape into lost dreams.

They called number ten, but Jean didn't react. They called his number a second time, and I had to elbow him out of his daze.

He lifted his head, moving forward mechanically, like a man condemned, marching toward his fate. It was hard to watch. Jean reached for the microphone, grabbed it, held on to it. He was exhausted. Finally, he unleashed a scream.

The single note, tense and bottomless, gripped the audience with dread. A song of anguish from time immemorial unfolded powerfully from Jean's blazing throat. From the shadows and darkness of an April night, Jean hooted into his dark forest. I recognized the owls conferring late at night in the valleys of our childhood. It was miraculous. I'd never heard such an imitation: Jean's vibrato was supernatural. A tawny owl had entered the competition, and, in silent flight, the raptor sought its prey. It called out, long and sustained, to the night. The song was so real that the audience froze, everyone holding their breath, terrified the hunter would see them.

Despite his cruelly changing voice, Jean had risen from the ashes of his herring gull. He beckoned the audience to follow him into the night-world. I watched his body and his breathing, and noticed he was singing on the inhale, not the exhale. At the start of each lament, he emptied his lungs to inhale sharply and trigger the sound. I saw his belly and sternum hollow, mimicking the gull's call in reverse, and opening up new horizons, new high notes. Jean's body was changing, and as it did, his birds were earning new wings.

Molting

BACKYARD EASTER-EGG HUNTS only last a few minutes, but they're one of the best childhood memories. I feel the same joy when I come across bird feathers on the ground. Sometimes I find remiges, those long, stiff flight feathers. There are also the rectrices, the flight feathers of the tail, which act as a kind of rudder, and the tectrices, or covert feathers, which protect the bird like slate roof tiles. When I come across a feather, I'm like a detective, trying to figure out what bird it might have belonged to. At the end of winter, birds discreetly lose countless fluffy, whitish feathers, which provide a layer of thermal insulation against the cold. Colored feathers, meanwhile, are the most sought after, but birds don't lose those until the end of the breeding season. I usually gather feathers one at a time, but now and again I come across a real treasure trove—an owl on the side of the road, or a chickadee half eaten by a sparrowhawk.

Molting is important for birds to replace and renew feathers damaged by time, ocean spray, or accidents. A feather is to a bird's plumage what a stone is to a cathedral—a basic element that, when combined with others, opens up all kinds of architectural possibilities. Feathers are one of the best examples of nature's genius, and such a source of wonder: they're both light and solid, they're waterproof, and then there are those iridescent colors... They're soft when you brush them in the direction of airflow, and rough against the grain. When you hold one between your thumb and your index finger and flick from top to bottom or blow hard, a feather will open to reveal its complex, repetitive architecture, with hundreds of barbs and barbules clinging together, infinitely multiplying to make a cohesive whole.

At the heart of the plumage are fine, barbless feathers called filoplumes, which look more like hairs, or like a cat's whiskers. Filoplumes provide the bird with constant information about the position of its wings and body in space. I remember on the farm when Monique would burn those hair feathers off chickens over the gas stove, and a sulfurous smell would fill the kitchen for hours—the smell of burning hair, which dragged us deep into some primitive pool of our common constitution as living beings. How can keratin, a sulfur-containing protein similar to human hair and to a lion's claws, assemble itself into such diverse, colorful structures, and fulfill so many different functions?

The feather is also the bird's Achilles' heel. The slightest mishap can be fatal. When a murre's feathers are coated in oil from an offshore spill, for instance, its life is in grave danger.

When you find a fledgling's feather, you can imagine the bird gradually molting, getting rid of its juvenile plumage, and growing up, one feather at a time.

Molting is also an outward manifestation of sexual maturity. Some species' immature plumage is neutral, which reduces adults' interest in younger birds and helps avoid aggressive or sexual behavior toward juveniles.

As years pass and adult feathers appear, birds' behavior changes slowly as well, as is the case for herring gulls. Juveniles are mottled brown, while second-year young are grayish; it takes about four years and six more or less complete molts for the adult coat to appear.

The feathers of some male birds often have a shimmery look; females, who sit on the eggs, would be easy prey, so a more muted, cryptic plumage helps camouflage them.

Among phalarope species—northern waders—the female is more colorful because after laying her eggs, she leaves the male to incubate the eggs and raise the chicks. Nature has so many adaptations and behaviors!

Just as young birds molt and change as they grow, I changed too: my voice broke. I don't remember exactly when it happened. My voice started cracking and deepening, like the tide coming into a misty bay—imperceptible, insidious, unstoppable. After a few warning signs, the wave crested. A few hairs around my mouth darkened, and my chest hurt. It would pass, the doctor said, it was only a phase. But the birds in my throat had left for good, a migration in the storm of hormones.

I had no shimmering feathers to show for it, no fancy crest on my head, no ornamental eyespots, no glimmering speculum patch in my plumage. I fought against this man I was becoming, a man I didn't want to be, not yet. I wanted to protect my voice, and the birds I didn't want to lose. But testosterone took its toll and my treble fizzled, derailed and awkward. It made people smirk, and was profoundly humiliating. I tried to work on other whistling techniques, but nothing worked.

Then, one day—I don't even know how the idea came to me—I finally found the solution: I used the airflow in the opposite direction to create a sound. Instead of exhaling, I inhaled, which dropped the vibration of my vocal cords by an octave. All the bird calls that open on an exhale are possible on inhale as well; all you need to do is tighten the muscles in your throat so as not to cough, since sucking in air through your mouth to produce a sound is a nasty feeling. The calls I managed were nearly the same as before, albeit slightly less intense, because exhaling is louder.

I was even able to improve some calls, like the tawny owl, with a vibrato I created by moving the base of my tongue, and the striking coo of the turtledove. My repertoire of birdsongs would remain.

As for my feather collection, I came across it not so long ago when I moved house. When I saw it, right away I wondered what budding ornithologist I might give it to. But when I opened the box, it no longer held a hundred specimens of flying species. There was only one, a skin beetle, an insect that feeds on animal matter, and which had turned my collection to feather dust.

Pecking Order

IT WAS THE START of a new school year. My parents had decided to bet on me as if I were a prize thoroughbred, and off I went to Saint-Pierre school in Abbeville, right next to the Hôtel de France. On the sage advice of teachers and friends, I had been switched from a public school to a private Catholic school. I was gifted academically, but far too easily influenced. I was drawn to slackers and nerds alike, as long as they were passionate about something—tractors, sports, history, theater, art… I joined the chess club and the handball team, and I represented the school in intercollegiate cross-country—a range of activities that made for both good friendships and bad company. I was like that boxer's son from Le Tréport who was a magnet for brats and bad behavior.

My parents thought the new school would motivate me. I had been admitted as a boarder at Saint-Pierre, which had sprawling green lawns and ten-foot-high walls. And with doctors' kids, lots of them, and whole families of lawyers' children. The crème de la crème gathered in the tiny attic rooms of the enormous old building in Abbeville.

Every time I went home the gap between school and the Arrest countryside got wider. I learned the codes of the bourgeoisie. I learned to argue, to defend myself with words. I lost the accent that signaled my social class. The boarding school wasn't an easy place for me: others looked down on me, and it was hard to fit in. I refused to talk about my parents' jobs or their tastes and passions; I didn't want to be laughed at. I even invented a father at the head of a major corporation. My clothes changed, and so did how much they cost. My parents were being bled dry to pay for boarding school. Needless to say, I kept my bird-calling talent a secret. I understood that soccer wasn't trendy and that bocce wasn't cool, let alone ornithology or a love of nature.

On our weekly half days off on Wednesdays, we would meet up in our classmates' huge mansions to talk about golf, swimming, and music. Every Friday afternoon, I waited for the bell to ring, my leather suitcase packed and ready. My mother and her little Renault waited a little ways away from the procession of fancy cars: in my foolish teenage pride, I made her stay out of sight.

Weekends in Arrest were tedious. There wasn't really anything to do. The only distraction was the bus shelter down the road, which doubled as an underage beer bar. I would head over with a whistle and a soccer ball, but no one wanted to play anymore, my childhood friends sinking slowly into the bottle. I didn't drink; I saved my gullet for the birds and my body for sports. I would end up staying in my room and doing my homework, listening to a turtledove on the neighbor's roof; just another Saturday.

One Sunday, my favorite team was knocked out of the Gambardella Cup by Amiens SC, and I slunk into class the next day, still stunned by the defeat and praying to somehow avoid afternoon math class. At four o'clock, the teacher arrived, fresh as a daisy and with a big smile on her face, no doubt delighted at the prospect of making us suffer. But instead of talking about our homework, to everyone's surprise, she asked me to stand. "You should all be proud to have Johnny Rasse in your class. Did you know that he can talk to birds? The principal told me; he knows the bird festival well. Johnny, can you give us a demonstration?"

I knew she was friendly with the principal. Without thinking, I deadpanned, "You should ask the principal; he really thinks he's the cock of the walk." The whole class burst out laughing, and the teacher's face went bright red. Coldly, she asked the class prefect to walk me to the office, to go repeat what I'd just said.

We waited outside the office for an hour, staring at the ceiling and at a copy of a Brueghel painting depicting the fall of Icarus. Finally, the door opened. Before the Louis XVI–style desk sat my mother.

"Can you explain why you were rude?" the principal asked. I didn't hesitate for an instant: "I didn't insult anyone, sir. The rooster is a sublime bird, with flamboyant plumage and a divine song." My mother elbowed me, but the principal shook his head. "No; go on."

I stood up and mimed all the barnyard characters—hens, turkeys, and guinea fowl, the whole lot—ending with the red junglefowl, a bird native to Southeast Asia, the ancestor of the barnyard rooster. Bird names weren't insults, I told him: I associated the role of the principal with the rooster because he was the one who was responsible for making sure the school aviary ran smoothly. Every teacher is a bird, I said: the French teacher, with her hooked nose and shrill voice, was a parakeet; the physics and chemistry teacher, who had three renegade tufts of hair at the back of his head, was a heron; and so on. I went through the entire teaching staff, though of course I didn't say a word about the rooster's sexual antics. The principal was impressed by my knowledge and my performance. He calmed down, promising my mother he would come see me at the next bird festival. He even walked me back to class, right across the schoolyard in the middle of recess.

Meanwhile, my reputation was made: all of a sudden I was the new kid who had insulted the principal in the middle of math class. That evening, all the seniors came to my dorm room. The event was recounted over and over again, with some embellishments. Thankfully, the inciting incident—the math teacher asking me to imitate a bird—seemed to have been completely forgotten.

I was relieved: the classroom was no place for bird calls. I wanted to keep that to the Abbeville theater. This time, I was convinced it would be my year. Jean had won the previous competition, so he wouldn't be in the running. The principal would be in the audience, proud as a peacock… or a rooster. And I would be the one on the front page of the *Courrier picard* holding the gannet trophy!

Bragging Rights

MY DREAM HAD FINALLY come true: I'd won the bird festival competition! I was finally holding the trophy in my arms, that little stone statuette depicting two majestic gannets, beak to beak, parading. This mating scene would actually have been impossible to ever witness in the Baie de Somme, since gannets only ever fish offshore, and if they are found on the beach, it's usually bad news environmentally. In any case, this was the icon that had been awarded to the winners in every category since the first edition of the festival: it was presented to Luc Jacquet for his film *March of the Penguins*, to Laurent Charbonnier for his documentary on deer, and now to me for my rendition of the northern lapwing. I was in prestigious company indeed.

But this wasn't the prize I wanted. For some time, I'd felt like I was at a standstill, or even getting worse. I still couldn't whistle with my fingers, a technique Johnny excelled at. And my voice was still playing tricks on me. I couldn't hold the long high notes anymore.

Luckily, that year's mandatory bird was the lapwing, a black-and-white bird whose song was just right for my voice at the time. They say the male's hoarse whistle as he twirls over his territory sounds like winnowing grain.

Before the mechanization of agriculture, wheat was shelled and winnowed by hand, usually in the colder months. Every yard rang out with the music of winnowers tossing seeds in the air in large baskets to separate the wheat from the chaff—a scuffling that sounded like the lapwing.

On the morning of the competition, I hoped my voice would hold, even if it tended to start breaking by midday. But the jury chose, and I won. The prize was a trip for two to Greece, which I gave my parents; in exchange, my sister and I would have a priceless week of peace and quiet. My victory also meant I would be on the jury the following year.

During the pre-competition dinner, one of the contestants, an old hand from down south, was talking about a bird-call tournament in the southwest. The competition was fierce, he told us, and the prizes were nothing like what we got in Abbeville. Down there, people said you could really win big—maybe even a car!

The man from the south said he'd come close to winning the previous year. Given his level, I would have a good chance. I did some research and discovered that the competition would take place on August 25 in Casteljaloux. All I had to do was negotiate a little side trip along the way on our family vacation. There was still a year to go, and I wasn't above a little bit of wheedling.

So it was that at the end of August, we arrived in a charming town in the Lot-et-Garonne region. Five hundred miles away from the Baie de Somme, I was completely lost. The accent was outrageous, especially that of little old men. All I got out of them was *palombe*, dove. Even the scenery seemed monotonous, lots of pine trees and ferns. I ran into the man from the previous year's competition in Abbeville, who looked half embarrassed and half pleased to see me.

The selection was open, and we could do any number of birds. There were two categories, *palombes* and everything else. I remember my first competition, where everyone imitated what we call the *oigne*, the wigeon. I guess every region has its official bird. What was unusual for me here was that contestants themselves announced the birds they were imitating. My turn came before I knew it. I was off: first, waders, sandpipers, plovers, and redshanks, then the two curlews, avocets, and passerines, from smallest to biggest, from the wren to the crow. Then ducks, and then the diurnal birds of prey—falcons, hawks, kites—and the nocturnal birds, the barn owl, tawny owl, little owl. Finally, the heron, and other species that force their call, like the cormorant and the grouse. My fatal mistake was ending with the common wood pigeon. The audience's collective reproach was instantaneous: "*Palombe*!" The judges sitting across from me were looking at me funny. Something was off. One of them stood up and argued with the organizer. They didn't know Baie de Somme birds. My fellow contestant from up north came to my rescue, explaining that I'd just won at the Abbeville festival, and that the competition here was much tougher, and the prize—more or less a trip around the world!—much bigger.

A short man in a beret won the pigeon competition, while I came first overall, an honor that was marked with a small silver trophy. I was expecting to win at least a car, but they handed me a box. I opened it; maybe there were keys inside. All I found, unfortunately, were three bottles of wine… and plonk at that, according to my father.

In October, I got a phone call from the Abbeville bird festival.

"We need a do-over!"

"What?"

"We need to film you again. TF1 shot the whole bird-calling competition last April for their natural history series. They have all the scenes, but they lost the last bit of footage and don't have the winner."

So there I was, back at the Abbeville movie theater. My parents, along with a dozen or so chamber of commerce employees, were seated in the auditorium, sitting in for the 850 people that had been in the original audience. They were supposed to clap to mark my victory, and the producers kept asking them to move around the room to look like a crowd, first at the back, then near the front row, like curlews rushing in to the mudflats when the tide goes out.

The third-place winner was onstage with me, as well as Olivier, a former winner who hadn't placed in the most recent competition but who'd been called up to fill in for the second-place winner.

I was in for a rough ride. I hadn't practiced in three months. I wouldn't be competing the following year, and it had destroyed me. I'd stopped training. The context was off, too: there was no draw, no dinner, no butterflies in my stomach before going up onstage, no prize, no hope of victory. Above all, there had been no birdsongs to practice. I didn't even know what the mandatory species was this year, while normally I'd have been pestering the organizers to find out as soon as possible that one bird that would keep me awake for weeks before the festival. There I was, standing around with ten extras, a mic, and three cameramen, and I had to perform a lapwing, which I hadn't done in months. I was horrible. No matter how hard I tried, nothing came out quite right. I suggested doing another bird, but no, this was the image they wanted, so I was forced to spit out a pathetic burp. It sounded more like I was ripping the lapwing's crest feathers off. The episode was screened over and over again, thankfully often at ungodly hours, though for a long time night owls up watching whatever was on television must have had a good laugh.

Number Eleven

IT HAD BEEN A WHILE since Jean had stopped by. He would stride past our house without a glance, head held high. Since he'd taken home that trophy, he seemed sure of himself—proud, almost haughty. When he left the theater after the last competition, he had told my parents that I'd made up my chaffinch entirely. He'd even asked me where I'd got it from. My mother, who rarely gets mad, grabbed my arm to avoid a confrontation. Jean had gone from being the *Miaule*, known for his herring gull, to the Boucault boy; my parents now just called him a smart-ass.

April had come around again, and there was no doubt in my mind that I would conquer both the audiences and the judges. I had chosen spectacular birds, the best singers with the highest-pitched songs, and the hardest to imitate, starting with the devastatingly powerful nightingale, a song I would perform with a new whistling technique. Then I would do the chaffinch for Jean, just to show that, yes, we did have some in the garden. I would wrap up with the blackbird, the required species in this competition, which had been with me since childhood, and which stood for perfection in my father's eyes.

On the day of the competition, I decided not to attend the dinner or to meet with the other contestants at the Hôtel de France. I avoided the parade through town, too; I dreaded the possibility of running into a high school classmate wandering around Abbeville on a Saturday night. I went straight to the theater, an hour early. The contestants were the usual suspects, sitting in the front row. As far as I was concerned, they were all pretty predictable; none of them stood a chance. There was the contestant from Mont-Saint-Michel with his throaty rasp, the men from the south with their pigeons and partridges, and the Baie de Somme guys with their wigeons, oystercatchers, and, of course, their *chés corlus.* I smiled inwardly as I imagined their struggle with the mandatory blackbird. I couldn't wait to find out how they were going to do it.

Out of twelve contestants, I drew number eleven—the best possible outcome. Denis Cheissoux, who was emceeing again, called my number. I stepped up to the microphone. I filled up my lungs and spread my arms like Christ on the cross. The nightingale's four long trebles soared up to the heavens. At the peak of the song, when I had to do a forceful and technically demanding warble, I attempted the longest, most powerful trill in the history of the festival, with jerky diphonic notes, in a whirlwind that lasted a solid minute—a record! When I finished, the entire audience jumped to their feet. Even Denis Cheissoux was overcome: "What breath, what singing… It's incredible!" Emcees were obviously strictly forbidden from weighing in, and as long as the festival had existed, none before had ever publicly given their opinion during the competition.

I watched the contestants fumble through the second round with standard oystercatchers and redshanks. The partridge and the wood pigeon got good laughs. Then it was my turn again with the chaffinch. I decided to whistle without my fingers, adopting another technique, the one Zorro usually used. The song is a crescendo, followed by a rapid decrescendo. I was thinking of Jean. He was on the jury that year, sitting behind the curtain (the judging was always blind). The purity of my singing and my execution speed suggested a sturdy chaffinch, sure of its charm. For the closing signature, I chose an option contrasting the three ritornellos. After a moment of utter silence, the audience broke into thunderous applause. As I returned to my seat, one of the other contestants breathlessly confided that he had never heard such a faithful imitation.

Then came the third round, with the blackbird, the obligatory species. You could see the anguish on everyone's faces. The contestants were all looking at each other, as if their competitors' eyes might hold the answer: *How are we going to do this one?* To my surprise, several contestants tried to escape the blackbird's artful warble by resorting to different calls: contact calls, flight calls, the call of nestlings... No one was brave enough to attempt those long treetop phrases. The end of the blackbird's song marks the outer limits of its territory, but it was as if tonight the blackbirds were lost, without a home or family.

When it was my turn, I stood in front of the microphone and imagined myself at the very top of the big cherry tree in the garden. I was at home here, and I was going to show the audience my range. The melody that emerged captivated the room, I could tell. The song of the blackbird—the most romantic of birds, the Frédéric Chopin of gardens, sharing a nocturne just as the sun goes down—tinged the present moment with longing for the past. My song faded gently into the shadows of the foliage on a blissful night.

The competition was over. A bunch of people came over to congratulate me. Some of them had been following me for a long time. The men loved the confident virility of the nightingale, while the women went on and on about the blackbird and its nostalgia for spring. Connoisseurs raved about my imitation of the chaffinch. Some wanted to take me out to dinner after the competition to have me perform for them, which seemed so strange to me. I caught Jean's eye, and he winked at me and gave me a thumbs-up. I knew that, even behind the curtain, he had recognized my songs and my birds. It was for him, too, that I sang so divinely that night.

After intermission, we were waiting for the final rankings, and the awards. Denis Cheissoux came in, smiling. He announced the results, starting with the final contestant, the twelfth, who was none other than the man who had come to congratulate me. Then Cheissoux announced that eleventh place went to… "Contestant number eleven." Silence. He went back to his notes to confirm.

"Yes, contestant number eleven, Johnny Rasse, is ranked eleventh." My April dream shattered into a thousand pieces.

The audience rustled in dismay. How could this be? They booed and fussed, but the decision was final. I had come second to last in a competition I was supposed to win. Gone were the headlines, the little statue of a gannet, the speech in which I would thank my parents… my poor, hopeful parents. I couldn't process what I was hearing, I couldn't smile. I lumbered onto the stage, stricken.

"Yep, some big surprises tonight," Denis Cheissoux said. It was the worst night of my life.

The rankings made no sense. I watched, stunned, until the winner was announced. It was the man from the south, who'd been a laughingstock for a decade with his doves and partridges, the braggart, with his swaying and thrusting that had the audience in stitches. I was so angry. I'd been robbed of my trophy, the whole competition, the victory.

When I finally found my parents, they patted my back and my head to comfort me. They whispered platitudes, trying to make me feel better despite the affront. My mother was so gentle. "It's okay," she shushed me. I couldn't look at my father. I heard him behind me: "I've got nothing to say. You were great, Jojo, it was sublime. Just don't let me run into Boucault, that arrogant prick. He had it in for you, it's all his fault!" My father was so mad that I didn't dare reply. I could almost feel my mother's concern that he would see Jean. As we left the theater, my father told anyone who would listen that it was over: I wasn't coming back, and the festival could shove it. As if we were the most important people in the world. My mother tugged him by his sleeve, back toward the car.

My father was right: that was my last Abbeville festival. In the car, on the way home, a huge weight lifted from my body. I felt so light. That much closer to the birds…

Behind the Curtain

WITH JUST A FEW days to go before the competition, I was finally starting to feel a little more comfortable with my role as judge. I found out what the mandatory species was: the blackbird. If the organizers had wanted Johnny to win, they couldn't have picked a better bird: according to his father, Johnny did it better than anyone else.

I remembered that my mother used to use grading scales when she marked French papers, and I set up a similar system for judging. I broke down each bird's song into four parts, each of which I would score out of five. For the blackbird, I obviously wanted to hear the male's territorial song, the dusk onset call, flight call, and alarm call, and, as the icing on the cake, the fledgling call. But my scale didn't work for species with fewer song variations, such as the barn owl; there, I could maybe award marks out of ten for the screech, and ten for the twittering of the owlets. I would have to tailor my scoring system to each species, to be as fair as possible and honor the birdsongs.

For the first time, I didn't head to the stage door; I went in through the foyer with the audience. Everyone was smiling at me, and I caught a glimpse of Johnny's parents waving. I was on the jury... Were they trying to influence me? This was a nice position to be in, I thought; I understood why Zorro had declared himself out of the competition for a few years now.

After the awards for the best nature films and bird photographs, it was time for the bird-call competition. When our names were called, each judge stood up and came onstage. I introduced myself. Applause. I could hear people calling out the names of different gulls, but I wasn't sure whether it was jeering or encouragement. The ten judges were invited to step behind the curtain that shielded the contestants from view. There were twelve this year, who had drawn numbers to determine their order. I probably knew most of them.

We were off to a good start with the first contestant: an oystercatcher, which was original. It wasn't very strong, just a call and a contact call, but no alarm call. Fourteen out of twenty. Contestants four and six also did a good job: a common redshank and a greenshank, the latter almost better than the former, but I listened with my eyes closed and a throat-slapping noise prevented me from imagining the bird. Eight out of twenty.

The seventh contestant I was positive was the fellow from the southwest that clinched my win last year. He did the red-legged partridge, which wasn't a bird I knew well. If it had been a gray partridge, from the plains here in the north, that would have been fine, but this bird was from the vineyards in the south. He did two different calls, which were average. I heard an animal that made me forget the human. But I found it hard to judge. For the originality of the bird, and when in doubt, I gave him a fifteen out of twenty.

When the eleventh contestant came up, even though I couldn't see the room, I could hear a strange silence settling, for a long time. Maybe he was adjusting the microphone to his height; he must be small. "The nightingale," the emcee announced. I heard a very loud whistle, much too loud. Johnny was signing his crime! He was the only one who would use his fingers to sing like a sparrow. We were treated to a minute of absolute genius. Clear, powerful, jerky notes, strung together so rapidly that it sounded like several birds. For the first time, the audience, which until then hadn't clapped for anyone, let themselves go. Applause was forbidden to avoid swaying the jury, which I think had something to do with me: my first bird dances drew rapturous ovations, and after that some grumpy contestants had asked for blind judging, without applause.

Yet Johnny's performance was an insult to the bird! He didn't hold any prefatory notes, and his phrasing was far too long. It was absolutely unrealistic. And what about the alarm call, the one that sounds like the common chiffchaff, only sharper, and the more surprising call that sounds like a toad croaking? Nothing! How can you claim to be a bird imitator and only perform part of the animal's repertoire? Sure, he did the greatest hits, and the prettiest songs, but there's more to a bird's life than seduction and mating. Each sound is meaningful for the animal. What makes a true bird caller is knowing those meanings and recreating them.

Five points for the melody, two for the contact call, zero for the alarm call and zero for the toad-like croak. Seven out of twenty. And I was being kind!

For the second round, Johnny had chosen the chaffinch. I heard a slight decrescendo, a sequence he repeated seven times. I would never achieve his level of technique, but I didn't hear a chaffinch in his imitation, not even the muffled, trilled alarm notes, which for some reason are dubbed the rain call. No female call, no contact call. Zero plus zero plus zero plus five, for a total of five out of twenty.

Sitting next to me was a ninth grader who had been selected at random to sit on the jury. I hadn't noticed, but he was probably worried about the accuracy of his judgment and had been copying my scores all along. My rankings would count for two, I realized to my dismay.

After the first two rounds, I'd already recognized all the contestants—Olivier, Pierre, Sébastien, and, of course, Johnny.

The last round was the mandatory species, the blackbird. The first contestant didn't stand a chance: all he could muster was a rough fluted sound for the melody, and an attempt at a contact call. Three out of twenty for effort. The finger whistlers did try for the melody, but they tended to forget the high notes at the end of the phrase, and none of them did an alarm call, that recognizable swirl, with the piercing high note, the shrill curl that never fails to make you look up to the sky, where often you see a circling hawk, which the blackbird has already spotted.

The contestant from the southwest was up: weak whistle, poor technique, an attempt at a contact, and the dusk onset and alarms calls were too low and hissed. Two plus four plus three plus two, for a total of eleven. I might have been generous, but he'd checked the boxes.

The verdict was the same for Johnny: his sequence was well whistled, but stereotypical. I knew that blackbird in the cherry tree, but this was too much. He overdid it, as always. His song was too long, too flawless. And, of course, he didn't do the chittering of the young, though I'm sure he would have been able to. Five, zero, zero, zero… five out of twenty.

The results were in. The guy from the southwest came first, which was surprising; it was the first time he'd ranked so high. His strategy of doing species with only one or two calls and birds that were less well-known here had paid off. Johnny was way down. He couldn't even place in the under-sixteen category, because he was two months too old. His father was going to be furious.

A month later, I appeared on *Qui est qui?*, the French version of a Welsh television show, which featured three contestants who were supposed to guess the profession or hobby of six people. Olivier was there as a bird caller, and I was a bird-call competition judge. The audience was expecting comic hijinks, fish out of water. Olivier may as well have been wearing a beak: the first contestant guessed what he was right away, before he even sang a note. Then he had to face me for the practical round, after which the audience would weigh in on who matched which profession. I suggested easy birds he could imitate almost better than me. Ninety-six percent of the audience got it. I came away from the experience with my first paycheck, and, at school, some new interest from girls. Two days later, the producers phoned me, hoping I might suggest another contestant. I gave them Johnny's number.

Gulls Under the Bridge

THE COMPETITION HAD BEEN RIGGED, my father was sure of it. But no matter how I looked at it, I couldn't imagine why Jean would have wanted me to lose. He'd always helped and supported me. The rivalry between us stemmed mainly from other people's perception of us—my father, Gilbert, the village of Arrest, the Abbeville audiences. Fortunately, after a week my father was in a good mood again: we'd just received a phone call from a TV show, *Qui est qui?*, hosted by Marie-Ange Nardi on Channel 2. The idea was to guess the profession or pastime of that day's invited guests. The producers would pay me and cover my travel expenses. My whole family knew about it: my uncles, grandparents, cousins… Everyone watched *Qui est qui.* After what had happened at the festival, I couldn't say no to my father. My only worry was that one of my school friends would happen to be watching and find out about my bird tendencies. That would be the end of me.

The family was proud again. We planned our trip to the TV studio in Paris. A friend of my father's, who assured us he knew Paris like the back of his hand, offered to take us. Thanks to him, we pulled in on set in the Plaine Saint-Denis suburb two hours late after driving around the Périphérique ring road three times. The studios were huge: ours was right next to where they shot game shows, like the French equivalents of *Going for Gold* and *Family Feud.*

I was taken in to makeup, my father standing behind me like he was my manager. He was right by my side every single second. When Marie-Ange Nardi whispered something in my ear, he wanted to know what she'd said, though he was disappointed when I explained that she was just making sure she was pronouncing my name right. He'd hoped it would be something about birds.

The game show began at last. Three contestants and the audience had to figure out who we were. They had been told there was an interpreter from Mandarin, a dog breeder, a canoeing instructor, a traditional African dancer, a kickboxing champion, and a bird caller. I almost passed for the kickboxer but gave myself away on a slightly technical question during the first round about how matches were organized. In the second, applied round, I sang the songs of the blackbird, redshank, and nightingale. Within two minutes, the show's audience record had been broken: by 96 percent, they pegged me as the bird caller. My father was thrilled, and our driver friend was charmed, though on the way home he confessed that he hadn't understood that they were supposed to vote, and was single-handedly responsible for the 4 percent who were off! I'd sure show them at next year's bird festival, my father said, laughing. I didn't say anything. There was no way I was going to enter the festival competition again.

The show was broadcast on a Wednesday. My mother pulled me out of school, claiming we had an important family meal. A party was organized at home. My grandparents were there, along with a few aunts, uncles, and cousins. I was happy to see them again, but I was still worried about how my classmates would react. When I came up on the screen, the whole family laughed, picking apart my haircut and my clothes. It was incredible how the camera magnified every little detail. My uncle asked me if Marie-Ange Nardi was nice, and my cousins wanted to know whether I had been nervous. My father, still playing manager, butted in, answering for me. They all laughed at the part where I didn't know the answer to the kickboxing question. When the moment came for the birdsongs, everyone fell silent. They were moved. The family celebration was a success. At the end of the evening, my grandfather said to me, prophetically, "You're going to do this for a living."

The next day, my stomach was in knots when I got to school. We spent the morning on a four-hour in-class English assignment. I rushed through the paper so I could escape before the rest of the class, and I avoided making eye contact with anyone until the last bell. I was saved: obviously, no one had watched the show.

On Friday night, back at home, I bumped into my brother, who was waving the remote control around. When he refused to hand it over, a fight broke out. I tackled him, and as we tussled, he pressed number four with his elbow—the paid Canal+ channel, which we didn't have, but the contents of which were available as highlights. Images of a plane crash in Florida flashed by, with hundreds of victims. Then I popped up on the screen, imitating a blackbird. A tragedy, followed by some kid whistling… I felt like a victim of Canal+'s sick sense of humor. I wanted to cry. They had made an ass out of me. And I knew the whole school watched that channel. All I could do was to fake being sick. I held out for a week, and missed an in-class biology assignment that was my makeup for the term, but it didn't matter; I had no other choice but to hide. And the plan worked: when I went back, none of the popular kids had picked up on the story, and I could finally resume my normal high school life.

It was mid-May. The chickadees were singing at the top of their lungs in the schoolyard. I bit my lip to keep from whistling. Sometimes, I would hide in the bathroom and sing, to make sure I still could. One time, the hall monitor, thinking she had heard a bird, was standing at the door with a broom when I came out. I couldn't wait for Fridays—climbing into my mother's Renault, leaving the city, the schoolyard, the bathrooms, everything. I couldn't wait to get back to my garden, and to the freedom to sing without having to hide.

That Friday, my father was already home, waiting for me with a smile on his face. He seemed happier than I'd ever seen him. As I was heading off to soccer practice, my father pulled me aside. He stared me straight in the eye, almost as if he were challenging me, and told me he had signed me up for Chambord.

"Chambord? Like, King François I?"

"No, in the Loire! You'll be the European champ!"

I didn't know what he was talking about, and it was up to my mother, once more, to clear things up. My father had entered me in the European bird-call championships, which would be held in a month, in the Chambord castle courtyard.

The European championships! Just hearing the words was a dream. That would show them, after the humiliation at the Abbeville festival. It was also an opportunity to take on other bird callers. And being crowned champion of Europe... For once, I agreed with my father.

A few days later, we got a package outlining the competition rules. The protocol was different than at Abbeville: first there were group stages, then knockouts, like in sports tournaments, and as contestants progressed through the stages, the number of birds to be imitated increased. There was one major restriction: contestants were to imitate waterfowl only, which excluded the birds I was best at, like passerines, blackbirds, nightingales, and finches. I was thinking of Jean and his irresistible gull: if he went to Chambord, and I didn't know if he was going, his herring gull would put to shame all the plovers, curlews, and pipers in the world.

In any case, this was my chance to get mine back after the disappointment in Abbeville.

But now I was stuck at boarding school in Saint-Pierre with no opportunity to practice! Seniors could avoid the cafeteria, but not juniors. Luckily, my mother arranged for me to visit my grandmother at lunchtime, which gave me two hours to practice on the banks of the Somme, working on the gull cry I was looking for. There was a good spot on the towpath, right next to a bridge where I could shelter in case of bad weather.

On the first day, I squawked and screeched under that bridge until two fishers came to beg me to stop: I was scaring away the fish! After a week of training, I'd completely lost my voice. I decided to take a break, but when I started up again, in the rain and under the bridge, I had to admit it: I couldn't do a gull for squat.

On my way back that day, I bumped into a friend, who was the drummer in a band. He had long rock and roll hair, Doc Martens on his feet, and earphones in his ears.

"Here," he said, handing me an earphone. "Listen, and breathe."

It was the Beatles' *Revolver* album, the last track, "Tomorrow Never Knows." From the first chords, the vibe was psychedelic. Suddenly, I looked up, intrigued by a sound.

"Herring gulls!"

"They're just seagulls, aren't they?" My friend was dismissive, but Lennon's gulls became a touchstone for me, and from that day on I met up with my drummer friend every day at break. I got to hear the Doors, Pink Floyd, and Led Zeppelin before he would let me hear the Beatles' gulls again. There was clearly the bird's rhythm and timbre, with very loud, low sections; the illusion was amazing. If the Beatles could imitate a gull's cry on a sitar and keyboards, surely I could do it with my own technique.

The next day, under the bridge, I had the rhythm and yawp of the gulls in my head. I no longer tried to shout. My fingers were my instrument. I found the right rhythm, but I lacked the cry's high range. I raised my head, aiming at the wall of the bridge. The acoustics under the bridge caused an echo, with two sounds rebounding simultaneously. I started in again on the high notes, and stopped short. That was it, that was the gull. I could do the cry of a herring gull with my fingers, I was sure of it.

After two hours of hyperventilating like that, my head was spinning. That afternoon, in class, I sat slumped in my chair, my face pale, and with a terrible headache. My friends and my French teacher thought I was on drugs.

Every day, I went back to the bridge. There on the towpath, the concrete arch created the resonance box for the diphonic call of the herring gull. All I had to do was somehow recreate the same echo chamber myself. I had to find a way to reproduce that telltale two-tone sound with my body. But how?

One day, while I was training, I came across the two fishers I'd seen the first time. Hearing me, they were speechless. One of them shouted in Picard for a third friend to come over, cupping his hands around his mouth to amplify his voice.

"*Y'a éne miaule sous ch'pont, tcho!*"

If a fisher thought there was really a gull under the bridge and even called others over to hear, my imitation must have been working. The third guy didn't know what was going on. "Get over here, I'm telling you," the first fisher repeated, again using his hands as a megaphone.

That's when it hit me. I'd always used two fingers of one hand to whistle, but if I cupped my other hand on top, I could create a double resonance chamber. That fisher's gesture was the key: I'd found the overtone I was looking for.

Chambord

WORD AROUND THE VILLAGE was that Johnny would be competing in the European championships. Olivier confirmed it: on June 21, a wetland birdsong imitation competition would be held at the Château de Chambord. He was going too. After what had happened with Johnny at the festival, it was obviously out of the question to ask him to take me along. The road test for my license wasn't for another month. Olivier agreed to take me, much to my mother's chagrin, since I had finals to study for.

She didn't want to let me go. "This competition will only upset him again," I overheard her lecturing my father. "This really isn't the time. And as soon as the written exams are over, you have to get ready for orals just in case!"

At noon on the last day of the written exams, as soon as I'd handed in my history paper, I ran home to meet Olivier, who was going to swing by to pick me up. I was so looking forward to the best test of all: a new birdsong competition. My bag was ready.

Two o'clock came and went. At three—no one. I started getting worried. At five, I phoned Olivier's wife. Something had come up, he couldn't go. From 5:03 to eight o'clock, I cried.

My father came home from work at the pharmacy, and I explained the situation: this was the European championship! I couldn't not go! And Johnny had already left. My father was on call from noon the next day until the following Monday, and couldn't very well abandon his pharmacy. I cursed Galen and Hippocrates; them and their oaths… "Unless we leave now," my father said. "I'll drop you off and drive straight back to work tomorrow at noon. And you can come home with Johnny."

I jumped up, relieved, and so happy. Although, given how his dad felt about me, it would be no picnic driving home with Johnny.

At dawn on the summer solstice, I found myself alone before the gates of the Château de Chambord while my father set off again on his return migration, like a pink-footed goose that rode the tailwinds and overshot its wintering site, then had to turn back in a headwind to find the right destination farther north. I was a little cold, and my stomach was gurgling with hunger, so I walked around the grounds. When I came back to the gates, two men were standing there. I recognized their silhouettes: it was Olivier and a friend of his. He'd played me.

Later in the morning, I met up with Johnny's family. They'd carpooled with Reynald Goldstein, a nature artist, who'd come to show some of his paintings of the Baie de Somme, though the area wasn't yet recognized for its raw natural beauty and sales weren't great.

Before the main event, the competition organizers had put together a quiz. They asked questions, and if participants had three correct answers, they got to take home a bottle of Loire wine, or a live duck. Name three wetlands in France: easy! Brenne, Camargue, and the Baie de Somme. Three around the world? The Danube delta, Djoudj in Senegal, and the Wadden Sea. Another duck in the bag.

I gave them all to Johnny's brother, Willy, and he brought them back to their father, who was starting to enjoy himself a little. By the end of the morning, we had a large crate with four hens and two drakes. We kept giving them little dishes of water, which they did nothing but spill. Our flock stopped there, when the wetlands game people finally banned me from the quiz.

Reynald had a tiny painting on display at his stand that caught my eye, of a little stream in the Baie de Somme, a tidal inlet surrounded by tall seagrass and swathed in mist off the sea. I almost didn't dare get too close: as soon as I saw it, I could smell the mud, hear the waders in the distance, almost feel the fog rolling in…

At ten o'clock, the first championship round began.

The rules seemed unusual. It wasn't a blind jury, for one. The judges were representatives of four different wetland conservation authorities. And the contestants, each of whom was assigned a number, had to draw lots to select the opponent they would face. This was a knockout match: you either won or you went home, with no wild card or double eliminations. It was a bit like the gauntlet for migrating storks: they had to make it across the Bosporus or the Strait of Gibraltar, or die trying.

There were about thirty contestants. A television crew was filming a little family, all of whom were taking part in the competition. They were picture-perfect, the parents and the three well-dressed blond children with straight white teeth; they looked like a commercial for laundry detergent or chocolate.

I tried to concentrate. I drew my focus inward. I didn't even look at what Johnny was doing. Finally, it was my turn. I drew my first-round opponent out of a hat: Olivier. My rage at having been tricked made my common shelduck transcendental, and my gray heron must have sounded more like a velociraptor. Olivier got out of there so fast he might have passed my father on the highway.

I dreaded having to face Johnny. But luck was on our side. In the following rounds, I drew each of the members of the telegenic family, who became less sweet-looking as we went on. The television guys were giving me dirty looks, but I couldn't help it, I had to make it to the next day!

It was five o'clock when the first round ended. Three of us were moving on to the final, which would be held the following day at noon in the square across from the castle, before an audience of thousands.

I went back to the little painting.

"Why is it the most expensive, when it's the smallest?"

"Because it's my favorite," Reynald replied. "And I don't want to sell it."

Two Birds

WE WERE FINALLY on our way to Chambord! We'd driven an insane 250 miles in one day. Reynald drove as he lived—in slow motion. He liked to take his time, chugging through little towns at fifteen miles per hour to make sure he hit all green lights. His paintings, which he hoped to sell at the championship, reeked of the varnish he'd just applied, so we had to drive with the windows rolled down and take a break every half hour—the smell was unbearable. But Reynald and my father were happy to discover the banks of the Loire, the city of Blois, its castle and its cave-like houses. Shortly before Chambord, we stopped on a bridge to stretch our legs. Rutilus roaches shimmered in the stream below, spawning. I thought of the bridge in Abbeville, and of my herring gull strategy, which my father still knew nothing about.

In the distance, a redshank was singing on a sandbank—a late departure for the season's migration. My brother heard it too. Without thinking, almost by instinct, I put my fingers to my lips and cupped my second hand over the first. My voice rang out, the call diphonic, full of attack. Reynald was startled and finally got into gear, running around, wide-eyed. My father thought a redshank had brushed by his head and almost fell over. He scanned the sky, not understanding. Both of them ducked down behind a low wall, as if to avoid being seen. My brother burst out laughing. Shut up, my father told him.

I launched into a second call, and the adults finally caught on. My brother stood proudly by my side. Just then, the redshank swept over us and landed on a lamppost on the bridge, with its beak wide open. It was completely surreal. The bird shrieked with pleasure—a cry absolutely identical to mine. Even from fifty feet away, it was impossible to tell the difference. My little brother ran to the lamppost. "I'm going to catch it!" He always wanted to catch everything.

Immediately, the redshank took off, and disappeared. My father, incredulous, asked me to do it again. I did. Again the redshank responded, from far away, but a few seconds later he flapped back over us and came to rest on the lamppost. The sun was on the banks of the Loire, a sublime evening. My father was mesmerized. "It's magic! The redshank is hypnotized! If you do this tomorrow, you'll win for sure. It's incredible!"

Our welcome was as memorable as the three days that followed would be. After another half an hour in the varnish-mobile, we arrived at the hotel. The drive had taken ten hours—twice what it normally should!

I got over my car sickness in a hurry when I saw the impressive, immense, majestic Château de Chambord. Enthroned in a verdant, forested estate, the castle looked like something out of my sister's favorite cartoons. At the reception desk, I learned that the competition was structured by elimination. Contestants drew each opponent live, in front of the jury. I would be there at ten o'clock for my first matchup.

I had time to stroll around with my brother before the competition got underway. We found Reynald's stand; his paintings were beautiful, but he wasn't really drawing a crowd. I saw the usual guys from the bird festival too—they were all here. There was a game stand, where Jean was answering questions. He seemed almost too happy to see us, as if he was a little lost by himself. I was careful not to say anything about the competition, but we both knew why we were here. He would be going after me, he told me. Jean and my brother decided to join forces in a quiz, to win ducks, and I went into the first round alone with my father.

At ten o'clock, I stepped into a booth to draw a number. I heard an Italian accent. The contestant I was about to face was quite old. He started, announcing his three birds: *anatra fischiante, pavoncella crestata, verde acqua*. I didn't understand a word. He took out three pieces of silver from his pocket and laid them on the table in front of him. This was what I had dreaded most: the poor little French boy was going to have to hold his own against one of those Italians with their damn mouth-whistles.

The man slid the first whistle into his mouth, and sucked violently. I recognized the song of the male wigeon, quite good, although it sounded a little tinny. It was my turn. For my first bird, I chose the redshank, which had welcomed me so warmly the day before. As soon as I made my first peep, the jury's eyes widened. It reminded me of Reynald's face, back on the banks of the Loire. I started the second run with my diphonics. The Italian jumped up, put his whistles back in his pocket, and, right in the middle of my performance, clasped me in his arms, shouting "*Campione, campione!*" He hoisted my arm up above my head like I'd just won a boxing match. I'd just knocked out the jury and my opponent at the same time. I guess this counted as a forfeit. The Italian had declared me the winner after a single bird! My father couldn't believe it, while I prayed that the Italian's prophecy would come true.

My next round was scheduled in two hours. In the meantime, Jean, my brother, and I roamed around, enjoying the various activities on offer at what was essentially a huge nature festival. We tried on beekeeping suits. We made bread. We built birdhouses. We won ducks.

The second round began. I drew another Italian, complete with sunglasses, a white polo shirt, and gleaming white teeth. I introduced myself to Ricardo. He pulled out a leather bag, inside of which was a small briefcase that looked like something out of a gangster movie. He had a collection of bird calls unlike anything I'd seen. I announced my list: common greenshank, Eurasian curlew, Eurasian wigeon. Behind his sunglasses, Ricardo was stalling, as if we were in a poker tournament, then presented his three—wigeon, greenshank, and curlew. The same birds as me! With a wide, snide grin, he sniffed, "*In bocca al lupo!*" I didn't understand. But I caught a glimpse of the old Italian from the first round, who winked at me with an encouraging thumbs-up. "*Campione*," he mouthed.

As soon as the jury opened this round, Ricardo raised his hand, like he had an emergency. He wanted to change his list. The judges conferred, and his request was granted. Ricardo reopened his magic briefcase. He announced the teal, pintail, and the greenshank, and, with the same sneer, he provocatively, sardonically wished me luck again: "*In bocca al lupo*."

I started off with the greenshank, that diphonic attack, before going into a mating call. Ricardo took off his glasses and placed them on the table. He closed his briefcase. When it was his turn, he did a teal duck that sounded more like a stationmaster's whistle. I went on with the curlew, which, with its trill, decrescendo, and alarm call, has one of the loveliest songs of all the waders. Rarely had I sung so well, I thought. Ricardo raised his hand. He wanted to switch birds again! After some consultation, the jury refused, and Ricardo picked up his briefcase and walked off, annoyed. Once more, a forfeit. I'd qualified for the semifinals!

My father was there. He wasn't saying anything anymore. Our eyes met. I could see the profound respect in his eyes. He understood that the man I was becoming was free, and would follow his own path.

While I was waiting for the semifinals in the afternoon, I bumped into Jean. He'd made it through all the rounds too. He'd gone up against French contestants without whistles, and I had eliminated Italians who used them. I thanked my lucky stars that we weren't slated to compete against each other for the time being.

The semifinalists were required to do five birds. In the draw, at two o'clock, I kept up my international streak: a Dutch fellow was picked to be my next competitor. The Italians I passed in the aisles addressed me as *campione*, which impressed Jean and my brother. I arrived at the table for my semifinal. My competitor was at least six feet tall, and red-faced. I'd initially mistaken him for a security guard. My head came up to his belly button, and he could have grabbed my skull like a handball.

I was doing the common greenshank, common redshank, Eurasian whimbrel, pied avocet, and Eurasian curlew. My opponent had a strange technique, with and without using bird calls. He had chosen the graylag goose, Eurasian wigeon, Eurasian oystercatcher, northern pintail, and black-tailed godwit. The giant started with the goose, honking huskily. It wasn't bad; a bit too serious for my taste. I did the common greenshank, using my new technique, impressing the audience once again. My opponent suddenly looked less ecstatic. His wigeon, for which he used a call, had less heart. I went on with the redshank, again attempting the mating call. The Dutchman tried an oystercatcher, but there was far too much air in it, and his fingers were too big. I followed up with the pied avocet; the audience was enchanted. Victory! I'd qualified for the final, which would take place in the courtyard in front of the castle, in front of several thousand people.

I checked on Jean. He would be in the final with me, as well as our friend from Mont-Saint-Michel. There would be three of us tomorrow, each of us doing five birds. I was thrilled: I was getting my revenge for that last festival.

Jean told us he was all alone in Chambord, with no family and no friends except us; he didn't even have a way to get home. My brother and I couldn't believe it, and set out to convince our father that Jean should stay at our hotel and drive home with us. Jean was full of surprises. At Reynald's stand, my father was talking to an Englishman and crying with laughter at the language barrier. They saw Jean, and my father asked him if he was in the final. "Yes," he replied shyly. My father was delighted. The Englishman didn't understand but went along with the general mood, repeating "*Bwavo, bwavo,*" like a parrot. "*C'est incroyable,*" my father said, and the Englishman repeated, "*Bwavo, incwoyabe!*" throwing his arms in the air. They looped their arms together and danced a little jig, swaying back and forth. The beer helped. We all laughed and danced together. My brother told my father that Jean and I had won ducks in the morning quiz game, and the dancing started up again. Jean was part of the family now.

The final was scheduled for the following day, so we went back to Blois. It was June 21, and the local mayor, Jack Lang, had founded a music festival that took place on the solstice. Everyone sang, shouted, and partied late into the night. I went to bed, leaving Jean and Reynald to paint the town. Tomorrow was a big day. The championship finals were like a big game: you had to prepare.

Music Fest

AT THE END OF the first day of the championships, Johnny's father told us it was time to leave.

"Come on, let's go to the hotel! Boucault, you're coming with us."

We left the castle, heading for the pension the Rasses had booked in Blois. We were right in the center of town. There was a music festival on, the weather was gorgeous, and everyone was out in the streets, which rang out with the sound of guitars and drums. We had to park a fair walk away and push through the crowd with our crate full of ducks. At the hotel, we snuck up the stairs to the second floor, hoping our host wouldn't spot our strange cargo. Thankfully, the room had a balcony. We put our ducks to bed. I eyed the small single bed, which I would have to share with Johnny's little brother, Willy.

We went out to have dinner. Reynald was euphoric: there was a potential customer who was interested, and who might come back the next day to buy a painting, so our painter friend treated us all to drinks. We ordered our meals, and the men had a large carafe of red wine. Naturally, they filled my glass too. I'd never had wine. I tasted it: it wasn't great, but it would do. I had an eel stew, which was delicious. The Somme was so polluted by PCBs that we couldn't eat the eels there. The carafe was drained and, somehow, miraculously, filled up again. After a few drinks and a lot of laughter, we got up to leave. Johnny's dad wanted to go walk around the music fest, and then we would head to bed. Tomorrow was a big day.

We headed down into the town. I was staggering around like a song thrush, the little bird whose Latin name, *Turdus philomelos*, at some point evolved into the French word for dizzy, *étourdi*, because the little songbird gorges on fermenting grapes, with predictable results.

There was music everywhere. We had to cross a bridge over the Loire River. The climb was interminable. I almost felt like swimming across. Night fell, and I could hear the thin screams of swifts. I looked up, but I had to close one eye to be able to see them at all clearly.

Johnny had had enough; he wanted to go back. His father chided him. Live a little, he said, words I decided to take to heart. My legs felt light now, like I was floating. I was so at ease with my new family. For a few more hours, we wandered around, stopping at various establishments to listen to different styles of music. Our musical selections were very eclectic, and mostly related to the length of the drinks line.

The journey back to the hotel was a little more demanding. The first staircase was carpeted, and a joy to negotiate, while the wooden spiral staircase offered the unexpected option of sliding down from top to bottom on my stomach. By the time I managed to get back on my feet, I was aching all over.

In the little bed, head to toe with Willy, I couldn't sleep. My ears were ringing, a dreary, high-pitched, agonizing whistle that sounded like the song of the greater hoopoe-lark on a loop. I didn't feel well. It was a terrible night: I would break out into a sweat, then shiver with cold. It was impossible to get to the shared bathroom on the landing without waking everybody up. And there was so much snoring.

Rematch

SUNDAY, TWO O'CLOCK, Château de Chambord. Thousands of people were gathered around a paddock where a show-jumping demonstration had just taken place. After some official speeches, the emcee announced the European bird-call championship finals. The guy from Mont-Saint-Michel, Philippe, was having a hard time. I was more determined than ever. No one was going to take this title away from me. Jean was focused but behaving oddly, and he was hanging on to his water bottle like a life preserver. We were in the middle of the courtyard. Unlike at the Abbeville festival, here the audience could express their opinion—loudly. The jury was right in front of us. My father had stationed himself right behind the judges' table, and was keeping a close eye on the score sheets.

Philippe dove right in, with a simple, effective oyster-catcher, the best I'd heard since the start of the competition. Jean opted for the common sandpiper, which was clever: his little whistling technique worked well with the mic. My turn. I asked to have the microphone moved back three feet, as a challenge, and launched into the redshank. It was all there: the territorial call, the mating call, the overtones. The applause was thunderous, and the emcee egged on the crowd. Philippe was stunned. Jean smiled at me. He'd just discovered my new two-handed technique, and must have been thinking I'd done a good job hiding my game. As far as I was concerned, the competition was over, but I tried to keep my mind on my birds. The debacle at the Abbeville festival still haunted me.

For our second birds, Philippe chose the redshank, and Jean, the shelduck. I was doing the curlew. Philippe had too much parasitic throatiness. Jean did a very good job, especially of the female. I was cocky, and allowed myself a few flight calls at the end of each melody. No one ever dared to try these, because they were so powerful. Some hunters hated them. Yet the flight call of the curlew was one of the most beautiful all along the French coast. I had a feeling that my new technique had widened the gap, but I couldn't feel the crowd's energy like in Abbeville. Here, the audience was fenced off, sixty feet away, though I did notice that every time I came up, people leaned forward.

For the next three rounds, Philippe went for the classics—a wigeon, a pintail, the European golden plover. He used the finger technique, but there was lots of mouth and throat noise, and I found his bird choices monotonous. Jean took on more complex birds, but his technique was limited. Shorebirds have to be heard across such a vast expanse, and over the noise of the surf, and their calls are almost all powerful whistles, which wasn't Jean's strong point, though the mic helped, and he managed some fine performances. As for me, I did the greenshank and the whimbrel.

The fifth, final round was the ultimate duel—what I had been hoping for since the festival. Jean's gull against mine. The cry against the diphonic whistle. But just as I was waiting to hear Jean's herring gull, the emcee announced… the curlew? I was dumbfounded. Why wasn't he doing his magic gull? I so wanted to see him charm the audience with that cry of his carried on the Loire wind.

It was my turn. When the emcee said "herring gull," Jean's face turned pale. He couldn't believe his ears. I went for it, alone, my arms stretched out. I rested one hand over the other. I was playing my last card; this was the moment I'd been waiting for. The illusion I'd created under the bridge in Abbeville was back again now here in the gardens of the Château de Chambord. When I finished, the audience started hollering. "Seagull! Seagull!" My fate was sealed.

I was unanimously declared European champion. Beyond the title, this marked the end of an apprenticeship, and the beginning of my liberation. I didn't need to talk about it with my father. He understood. I took a vow of silence for five long years. I let myself be drawn into the whirlwind of life and school. And, if ever I did return to the stage, I wasn't going to do it alone. That was out of the question. I would only ever do this again with Jean—the only real bird-man I knew, a rare, unique species.

The Morning After

WE GOT UP AT seven thirty to head to the castle. I caught sight of myself in the mirror: I looked like death. The drive was rough. There seemed to be a nuthatch on my head, tap-tapping on my temples at regular intervals. As a bonus, the sour smell of duck droppings wafted through the car from the trunk. I just wanted to get this over with. Reynald, to whom I'd confessed my hangover, told me to drink water, and I guzzled it, perhaps in keeping with the waterfowl theme of the weekend. We still hadn't discussed our species for the day. Every time I tried to broach the subject with Johnny, his father cut me off: "You'll see."

We came into the main courtyard, where long tables had been set up. A meal was planned, and we were the main course! A man came out, microphone in hand, to address the crowd. I couldn't see a microphone stand, which was better for me, since I whistled without my fingers and would still be able to hold the mic; Johnny was in trouble, though, since he wouldn't be able to use it.

It took forever to get going. First there were speeches by various officials: the mayor, the deputy, the senator, the sub-prefect, the prefect. We'd been out in the beating sun for a good forty minutes by then, and I'd drunk so much water that I needed to go to the bathroom, but I couldn't very well leave the courtyard. I was caged, condemned. I envied our ducks, which could happily relieve themselves in their crate.

At last, it was our turn. It would all come down to five birds. We had to enter our lists on a form and hand it to the emcee. Philippe went first. He whistled with his fingers too, so the emcee held the microphone in front of him, but much too close. His imitations were fine, but the mic picked up a glottal sound that spoiled the calls. Then it was my turn. My lips weren't working very well this morning, but the sound came through nicely.

I felt pretty good about my avocet and my common sandpiper. I only made simple sounds that I mastered easily. My shelduck was acceptable, but I couldn't use as much force as I would have otherwise, for fear of wetting my pants. Although storks and condors practice urohidrosis, defecating on their legs as a cooling mechanism, I didn't think the same behavior would go over very well coming from me.

Johnny came out. I'd never seen him so determined. His first four birds were very good; we'd be neck and neck for sure. But when the emcee announced his final bird, I felt like I'd been shot through the heart: the herring gull! What?! That was me, I was the gull!

Slowly, Johnny put his fingers in his mouth, covered them with his other hand, and began the gull's cry. The illusion was hypnotic. He didn't physically embody the bird, but I felt that the audience was experiencing something otherworldly. Something I had lost.

The results were as expected: Johnny was declared the European champion, and I was runner-up. Two thousand francs for him, a thousand for me.

We headed back home to Picardy. The barley and the wheat were still green.

When we stopped for a quick break, Reynald took me aside.

"Do you still like the painting?"

"Well, yes, but it's way too expensive."

"You know, yesterday, I said I had a customer who was interested. That customer was you. A thousand francs will do."

Dream a Little Dream

JOHNNY DISAPPEARED FROM THE competition circuit over the next few years. I continued on my own.

I offered to lead guided tours during the bird festival. The first outing was scheduled for a Saturday in April. I bought a gâteau battu and some apple juice, and packed my binoculars, my scope, and a few pairs of boots in case some of the participants didn't know what a marsh actually was. The festival phoned: no one had signed up. The second time, I had a little flock. Then, as word of mouth made my reputation as a naturalist, more and more people came out. It's so wonderful to discover the life of birds and to reveal it to those who don't see it, or who have forgotten how to see it.

Like a shepherd leading his sheep out to idyllic pastures, I chose secret paths and led my groups out to see rare birds. Until now, I'd lived these moments only for myself, but sharing my complicity with nature with others brought me intense joy. I was happy to be alive.

"Johnny has been trying to get ahold of you," my father said to me one day. "He came by the pharmacy yesterday."

It must have been five years since Chambord. During the week, I was a student in the city, and on weekends, my boots and I tromped around the bay. He wasn't doing bird-call competitions anymore; things felt easy between us.

"Yeah? What did he want?"

"I gave him our house number. He'll give you a call."

My parents had upended our lives. They'd sold the pharmacy in Arrest, and we now lived in Amiens. When Johnny came over, he'd changed, he'd grown up. He wasn't the same little boy he'd been five years ago. He was in school, here, on scholarship, he told me. Then he got down to business.

"I screwed up, Jean. I was driving my parents' car, and I ran a stop sign. At Saint-Val, you know, the stop sign at the scrapyard."

"Right! On the way down. That's an easy one to miss."

"It is. I got caught. I don't have the money for the fine. I'll be working at a summer camp in July, but right now I'm broke. So I asked my dad. He said he would pay my fine, on one condition: 'I want you to do the Abbeville bird-call competition one last time.' So I need you to help me. I want to do the duet with you. This year, it's the garganey. If we do it together, we'll win for sure."

The garganey, *Spatula querquedula*, is a small duck that returns from Africa to the Baie de Somme in early March. Locals nickname it "the cricket." The call of the male is complicated. Honestly, I was pretty pleased that Johnny would come looking for me, and that he thought we could win.

"Okay, you do the male and I'll do the female. The male is almost impossible. But it's all about the female, and I can do the female quite well. You know how the mating dance goes, right? The male circles the female and raises his body above the water, then he throws his head back and sings. You do the same, blowing from the back of your throat. If you try to imitate the beginning of a capercaillie, and then you do a mistle thrush, the two mixed together are like the male garganey."

That April, we won the duet hands down. The fine was paid.

The festival wasn't thrilled about our comeback: if the same people always win, other contestants get discouraged. The organizers were looking for some new gimmick, some way to shuffle the deck. The Abbeville bird festival's great bird-call competition was no more, they declared; it would henceforth be replaced by a mammal imitation contest!

I told Johnny.

"Are you serious? They didn't..."

"I'm telling you. Wolf, deer, stag, wild boar... And you're supposed to imitate one of the species, as a duet."

"That's nuts! I should warn you, I don't do mammals."

"It'll be fun, you'll see. I've been practicing. I can already do a pretty good wolf."

When the next festival rolled around, there were only a handful of us competing. I drew the first spot, and started with the wild boar. One evening, I had set out on the dunes at Marquenterre to try to hear nightjars, and found myself surrounded by a herd of boars. It was impressive, especially at night. I could hear the little squeals of piglets and the scuffle of the animals foraging. Suddenly, the sow picked up my scent and let out a frightened cry. The babies started running off in every direction around me.

I stood at the mic and replayed the scene. The theater filled with laughter, the same kind of chortling we would hear when a contestant bungled an imitation, slightly mocking—a sound I'd never heard before for my bird calls.

When Johnny went up to bellow like a stag, the same laughter echoed through the room. The tension we used to be able to weave between birds and humans, the magic we had created, had vanished.

The competition was over quickly. There were so few of us that Johnny and I waited in a little dressing room. We had no desire to face the audience.

"Did you see what we just did?" Johnny asked me.

"Yeah… I think we might be able to win."

"Who cares about that? You've been whistling like a god for twelve years, thousands of people clamoring to watch you be a bird, and now you're happy to just clown around, to make a fool of yourself in this circus? We've got to stop doing these competitions, Jean. People want to see us onstage. So let's do it: if we make it happen, they'll come see us, they'll come out everywhere." He paused. "What do you dream about, Jean?"

"Being able to see birds of the Amazon rainforest, and albatrosses."

"So let's go! We can even go to Japan, and Australia. We'll put on birdsong shows, and everyone will come. We can call ourselves…"

"… The Bird Singers!"

Epilogue

THE BOOK YOU HOLD in your hands tells the story of how our duo, Les Chanteurs d'Oiseaux—The Bird Singers—came to be.

Our career has been an extraordinary artistic adventure. We've had the chance to work with classical and jazz musicians, and to perform on renowned stages across France, including at the Arsenal in Metz, the Grand Théâtre de Provence in Aix, La Criée in Marseille, and the Lille and Paris opera houses; we've taken our act to the Théâtre du Châtelet, the Théâtre des Champs-Élysées, the Théâtre de la Colline, the Paris Philharmonic Hall...

Birdsongs are universal, and we're lucky to have played to delighted audiences around the world, chirping and warbling in Tokyo, Yekaterinburg, New York, Montevideo, Lima, Marrakesh, Warsaw, and Delphi. We also recorded an album, the popular *La Symphonie des Oiseaux* ("Bird Symphony"), for bird-men with violin and piano.

Twenty years later, those two boys who loved birds are still sharing their passion, with regular theater shows and outdoor concerts. And after the curtain closes, we always make our way back to the Baie de Somme, where it all began.

Stay tuned: who knows what's next in the epic, unique tale of The Bird Singers.

ABOUT THE AUTHORS

JEAN BOUCAULT AND JOHNNY RASSE first learned to imitate bird calls as children in northern France. Their rivalry at bird-calling competitions gave way to friendship, then to a partnership as performers under the name Les Chanteurs d'Oiseaux, or The Bird Singers. Although Boucault is trained as a pharmacist and Rasse as an engineer, their passion for birds has led them to create numerous stage shows, record albums, and tour throughout France and internationally.

For more information on The Bird Singers, videos, and tour dates, visit their website at chanteurs-oiseaux.com.

ABOUT THE TRANSLATOR

KATIA GRUBISIC is a writer, editor, and translator. Her work has won the Gerald Lampert award, the Cole Foundation Prize for Translation, and the Governor General's Award for translation. She has also been short-listed for the A. M. Klein Prize and twice for the Governor General's Award for translation.